Ethics for Engineers

Ethics for Engineers

Martin Peterson
Texas A&M University

New York   Oxford
OXFORD UNIVERSITY PRESS

Oxford University Press is a department of the University of Oxford.
It furthers the University's objective of excellence in research, scholarship,
and education by publishing worldwide. Oxford is a registered trade mark of
Oxford University Press in the UK and certain other countries.

Published in the United States of America by Oxford University Press
198 Madison Avenue, New York, NY 10016, United States of America.

© 2020 by Oxford University Press

Library of Congress Cataloging-in-Publication Data

Names: Peterson, Martin, 1975- author.
Title: Ethics for engineers / Martin Peterson.
Description: New York : Oxford University Press, 2019. | Includes
    bibliographical references.
Identifiers: LCCN 2018051680 (print) | LCCN 2019006980 (ebook) |
    ISBN 9780190609207 (Ebook) | ISBN 9780190609191 (pbk.)
Subjects: LCSH: Engineering ethics.
Classification: LCC TA157 (ebook) | LCC TA157 .P4285 2019 (print) |
    DDC 174/.962—dc23
LC record available at https://lccn.loc.gov/2018051680

9 8 7 6 5 4 3 2 1

Printed by Sheridan Books, Inc., United States of America

TABLE OF CONTENTS

LIST OF CASES

This book seeks to introduce the reader to some of the most intricate ethical issues faced by engineers and other technical experts. My ambition is to present the ethics of engineering in a manner that enables students to articulate their own moral outlook at the same time as they learn to apply ethical theories and become familiar with professional codes of ethics.

Over the past twenty years I have taught engineering ethics in three countries: the United States, the Netherlands, and Sweden. The topics covered in all my courses have been roughly the same; so even though this book is primarily written for the American market, the material is likely to be relevant in many parts of the world. As is customary in textbooks, I do not advocate any particular position on issues in which there is substantial disagreement, but I do try to make the reader familiar with what I believe to be the best arguments for and against the major views on controversial topics.

Some of the forty case studies discussed in the "boxes" are classic cases that have shaped the discipline for decades. Others originate from the course in engineering ethics my colleagues and I teach at Texas A&M University. Although the factual aspects of most cases are widely known, I have tried to focus on what I believe to be the most interesting *ethical* aspects. Technical details can sometimes be crucial, but it is important not to lose sight of the woods for the trees. Instructors looking for additional cases may wish to familiarize themselves with the collection of almost five hundred cases analyzed by the National Society of Professional Engineers' (NSPE) Board of Ethical Review over the past six decades. Many of the Board's analyses, which are based on the NSPE Code of Ethics, are very instructive.

A few sections of the book summarize ideas I have discussed elsewhere. For instance, some of the material in chapter 5 draws on a text on utilitarianism originally written for a Dutch textbook, *Basisboek ethiek* (Amsterdam: Boom 2014, ed. van Hees et al.); and the final section of chapter 4 is based on my book, *The Ethics of Technology: A Geometric Analysis of Five Moral Principles* (OUP 2017). Ideas from some previously published research papers also figure briefly in other chapters.

ACKNOWLEDGMENTS

I would like to express my sincere thanks to students and colleagues at Texas A&M University, Eindhoven University of Technology in the Netherlands, and the Royal Institute of Technology in Sweden for encouraging me to develop the material presented in this book. My co-instructors and teaching assistants at Texas A&M University deserve special thanks for allowing me to try out some of the pedagogical strategies outlined here in our large class in engineering ethics.

I would also like to thank my current and former colleagues for introducing me to some of the case studies presented here, as well as for challenging my perspective on many of the theoretical issues discussed in the book. Debates on engineering ethics can benefit enormously from discussions of real-world cases *and* ethical theories.

For helpful comments on earlier drafts, I am deeply indebted to Neelke Doorn, Barbro Fröding, William Jordan, Glen Miller, Per Sandin, Brit Shields, James Stieb, Diana Yarzagaray, and six anonymous reviewers. I would also like to thank my assistant, Rob Reed, for helping me to edit and prepare the final version of the entire manuscript. Finally, I would like to thank Robert Miller and his colleagues at Oxford University Press for their invaluable support throughout the production process.

Texas, July 2018

PART

What Is Engineering Ethics?

CHAPTER I

Introduction

Engineers design and create technological systems affecting billions of people. This power to change the world comes with ethical obligations. According to the National Society of Professional Engineers (NSPE), "engineers must perform under a standard of professional behavior that requires adherence to the highest principles of ethical conduct."[1] Other professional organizations use similar language for articulating the ethical obligations of engineers. For instance, the Institute of Electrical and Electronics Engineers (IEEE), the world's largest professional technical organization with members in 160 countries, writes in its code of ethics that "We, the members of the IEEE, . . . commit ourselves to the highest ethical and professional conduct."[2]

It is uncontroversial to maintain that, under ordinary circumstances, engineers shall not lie to clients; steal ideas from colleagues; jeopardize the safety, health, and welfare of the public; break the law; or perform acts that are widely considered to be unethical in other ways. However, engineers sometimes face ethical issues that are remarkably complicated and cannot be easily resolved by applying straightforward, common-sense rules. The crisis at the Citicorp Center in Manhattan is an instructive illustration (See Case 1-1 on the next page). Although this case occurred many years ago, the ethical issues faced by William LeMessurier, the structural engineer at the center of the crisis, are as relevant today as ever.

Familiarize yourself with Case 1-1 and try to formulate comprehensive answers to the following questions: Was it morally wrong of LeMessurier to not inform the public about the potential disaster? Did he take all the actions he was morally required to take?

The Crisis at the Citicorp Center

When the Citicorp Center opened on Lexington Avenue in New York in 1977, it was the seventh tallest building in the world. Today, more than forty years later, it is still one of the ten tallest skyscrapers in the city with a height of 915 feet and 59 floors.

St. Peter's Evangelical Lutheran Church originally owned the Lexington Avenue lot. The Church refused to sell to Citicorp but eventually agreed to let the bank construct its new headquarters on its lot on the condition that they also built a new church for the parish next to the skyscraper. To maximize floor space, William LeMessurier (1926–2007), the structural engineer in charge of the design of the Citicorp Center, came up with the innovative idea of placing the building on nine-story-high "stilts" and build the church beneath one of its corners (see Figure 1.1). To ensure that the load-bearing columns would not pass through the church beneath, they were placed in the middle of each side of the tower (see Figure 1.2). LeMessurier's groundbreaking design was less stable than traditional ones in which the columns are located in the corners. To understand why, imagine that you are sitting on a chair in which the legs are placed in the middle of each side instead of the corners.

Around the time the Citicorp Center opened, Diane Hartley was working on her undergraduate thesis in civil engineering at Princeton University. Hartley studied the plans for the building and noticed that it appeared to be more likely to topple in a hurricane than traditional skyscrapers. Although Hartley was not certain her calculations were correct, she decided to contact LeMessurier's office. Design engineer Joel Weinstein answered Hartley's phone call and assured her that her fears were groundless. However, when LeMessurier heard about the unusual phone call he decided to double check the calculations. To his surprise, he found that Hartley was right. The calculations had been based on the false assumption that the strongest wind loads would come from perpendicular winds. LeMessurier now realized that the unusual location of the load-bearing columns made the tower

Figure 1.1
To maximize floor space, William LeMessurier placed the Citicorp tower on nine-story-high "stilts."
Source: The photo is in the public domain.

significantly more vulnerable to quartering winds, that is, to winds hitting the tower at an angle of 45 degrees.

LeMessurier also discovered that the load-bearing columns had been bolted together instead of welded as specified in his drawings. Under normal conditions, this would have been an acceptable modification; but because the bolted joints were considerably weaker, and because the wind loads were higher than originally anticipated, the situation was now critical. After analyzing the situation carefully, LeMessurier concluded that if a

Figure 1.2
The new St. Peter's Evangelical Lutheran Church is located behind the traffic light in the foreground.
Source: Dreamstime.

sixteen-year storm (i.e., a storm that occurs on average every sixteen years) were to hit the tower, then it would topple if the electrically powered tuned mass damper on the roof stopped working in a power fallout. LeMessurier also estimated that winds capable of taking down the tower even when the tuned mass damper worked as specified would hit the building every fifty-five years on average. The probability that the tower would topple in any given year was, thus, roughly equal to the probability that the king of spades would be drawn from a regular deck of cards.

LeMessurier became aware of all this at the beginning of the hurricane season in 1978. At that time, the New York building code only required designers to consider perpendicular winds; so in a

strict legal sense, he had not violated the building code. However, LeMessurier was convinced that as the structural engineer responsible for the design, he had a *moral obligation* to protect the safety of those working in the tower. He also realized that his career as a structural engineer might be over if it became publicly known that the tower could topple in a moderate hurricane.

Who should be informed about the problem with the wind-bracing system? Did the public have the right to know that the building was unsafe? Was LeMessurier the only person morally obliged to remedy the problem, or did he share that responsibility with others?

The ethical questions LeMessurier asked himself in the summer of 1978 are timeless and equally relevant for today's engineers. After thinking carefully about how to proceed, LeMessurier decided to contact the senior management of Citicorp to explain the possible consequences of the problem with the tower's structural design. Before the meeting, LeMessurier worked out a detailed plan for fixing the problem. His solution, which Citicorp immediately accepted, was to reinforce the structure by welding thick steel plates to each of the two hundred bolted joints of the load-carrying columns. The welding started at the beginning of August and took place at night when the tower was empty.

LeMessurier also took several other measures intended to prevent a looming disaster. He informed the City of New York and the Office of Disaster Management about the situation. Together they worked out a plan for evacuating all buildings within a ten-block radius of the site with the help of up to two thousand Red Cross volunteers. LeMessurier also hired two independent teams of meteorologists to monitor the weather. Both teams were instructed to issue wind forecasts four times a day and warn LeMessurier in time so they could evacuate the building if strong winds were approaching.

On September 1, Hurricane Ella was approaching New York. LeMessurier's meteorologists warned him that the wind speed would almost certainly exceed the critical limit if the hurricane continued along its predicted path. However, just

(Continued)

as LeMessurier was on the verge of evacuating the tower, the hurricane luckily changed its path a few hours from Manhattan.

The public and Citicorp employees working in the building were kept in the dark about the true purpose of the nightly welding jobs, which were completed in October. Neither LeMessurier nor Citicorp felt it was necessary to reveal the truth to those not directly involved in the decision-making process. The full story about the crisis in the Citicorp Center was not revealed until seventeen years later, in 1995, when *The New Yorker* ran a lengthy article by Joe Morgenstern. He describes the press release issued by Citicorp and LeMessurier as follows: "In language as bland as a loan officer's wardrobe, the three-paragraph document said unnamed 'engineers who designed the building' had recommended that 'certain of the connections in Citicorp Center's wind-bracing system be strengthened through additional welding.' The engineers, the press release added, 'have assured us that there is no danger.'"[3]

Did LeMessurier (see Case 1-1) do anything morally wrong? According to the NSPE code of ethics, engineers shall "issue public statements only in an objective and truthful manner."[4] This seems to be a reasonable requirement, at least in most situations, but the press release issued by Citicorp and LeMessurier clearly mislead the public. The claim that "there is no danger" was an outright lie. This suggests that it was wrong to lie to the public. However, one could also argue that it was *morally right* of LeMessurier to mislead the public and employees working in the building. By keeping the safety concerns within his team, LeMessurier prevented panic and unnecessary distress. This might have been instrumental for keeping the public safe, which is a key obligation emphasized in nearly every professional code of ethics, including the NSPE code. According to this line of reasoning, preventing panic outweighed telling the truth. Of all the alternatives open to LeMessurier, the best option was to fix the problem without bringing thousands of people working in the area to fear a collapsing skyscraper.

We could compare LeMessurier's predicament with that of an airline pilot preparing his passengers for an emergency landing. Sometimes it may be counterproductive to reveal the truth to the passengers. Their safety might be better protected by just giving precise instructions about how to behave, meaning that it might, in some (rare) situations, be morally acceptable to lie or deceive passengers or other members of the public.

The take-home message of these introductory remarks is that engineers sometimes face morally complex decisions. Not all of them can be resolved by appealing to common sense. Although many of LeMessurier's actions seem to have been morally praiseworthy, his decision to lie to the public was controversial. It *might* have been morally right to do so, but to formulate an informed opinion, we need to learn a bit more about applied and professional ethics. By studying ethical concepts and theories more closely, we can improve our ability to analyze complex real-world cases and eventually provide a more nuanced analysis of LeMessurier's actions.

THE ENGINEERING PROFESSION

On a very general level, an engineer is someone who applies science and math for solving real-world problems that matter to us, the members of society. William LeMessurier clearly meets this characterization. A more formal definition, taken from the dictionary, is that an engineer is someone who applies science and math in ways that "make the properties of matter and the sources of energy in nature useful to people."[5] According to this definition, it is primarily (but perhaps not exclusively and universally) the focus on *useful applications* that distinguishes engineers from scientists. The latter occasionally study matter and energy for its own sake, but engineers by trade seek applications.

Having said that, other professionals, such as doctors, also apply science and math for solving real-world problems; and some engineers base their work on approximate rules of thumb rather than math and science. This shows that it is no easy to task to give an uncontroversial definition of engineering. A radically different approach could be to say that an engineer is someone who has been awarded a certain type of academic degree, or is recognized by others as an engineer.

In the United States, the practice of engineering is regulated at the state level. Each state has its own licensure board, and the licensure requirements vary somewhat from state to state. However, candidates typically need to take a written test and gain some work experience. Once the engineer has obtained his or her license, he or she has the right to use the title *Professional Engineer* or *PE*. Only licensed engineers are authorized to offer engineering services to the public and to sign and seal construction plans and other key documents.

Perhaps surprisingly, no more than 20 percent of those who graduate with a BS in engineering in the United States become licensed engineers. There are two significant factors here. First, engineers working for the government are exempted from the requirement to be licensed. Second, the so called "industry exemption" permits private engineering firms to employ unlicensed engineers to research, develop, design, fabricate, and assemble products as long as they "do not offer their services to the general public in exchange for compensation."[6] Because of this exemption, a private engineering firm can in theory employ only one licensed P.E. to review and sign all the company's plans. As the law is written, this single P.E. can be the individual who officially offers the company's engineering services to the public.

The regulations in Canada resemble those in the United States; but in other countries, the rules are very different. In countries such as the United Kingdom, Germany, and Sweden, anyone can set up an engineering business, regardless of education and actual competence. The absence of regulations is to some extent mitigated by university degrees in engineering and membership in professional organizations. Such pieces of information can help potential customers understand if the person offering engineering services has the necessary knowledge and experience.

Although the regulations engineers must follow vary from country to country, one thing that does not vary is the impact engineers have, and have had for centuries, on society. Railways, electricity, telephones, space rockets, computers, and medical x-ray machines have transformed the lives of billions of people. The list of famous engineers inventing these and other revolutionizing technologies is long and growing. Almost everyone in the United States has heard of John Ericsson's (1803–1889) propeller, Alexander Graham Bell's (1847–1922) telephone, and Thomas Edison's (1847–1931) lightbulb. In the United Kingdom, Isambard Kingdom Brunel (1806–1859) designed bridges (see Figure 1.3), tunnels, ships and railways that radically transformed the transportation system. In the Netherlands, the influence of the Dutch "superstar

Figure 1.3

The Clifton Suspension Bridge in Bristol in South West England was designed by Isambard Kingdom Brunel in 1831. With a total length of 1,352 ft., it was the longest suspension bridge in the world when it opened after 33 years of construction in 1864. Source: iStock by Getty Images.

engineer" Cornelis Lely (1854–1929) was so great that one of the country's major cities was named in his honor after his death: Lelystad. In more recent years, engineers like Steve Jobs (1955–2011) and Elon Musk (b. 1971) have transformed their industries in fundamental ways.

While the history of engineering is dominated by male engineers, revolutionary contributions have also been made by female engineers. Mary Anderson (1866–1953) invented and obtained a patent for automatic windshield wipers for cars. In 1922, Cadillac became the first car manufacturer to adopt her wipers as standard equipment. Emily Roebling (1803–1903) played a crucial role in the construction of the Brooklyn Bridge in New York, and Martha J. Coston (1826–1904) made a fortune by inventing a type of signal flare still used by the US Navy. More recently, numerous women have made it to the top of multinational tech corporations such as Hewlett-Packard, Google, Facebook, and Yahoo.

Having said that, it is worth keeping in mind that the vast majority of engineering graduates never become international superstars, although nearly all make important contributions to society, including those who do not have the luck, personality, or technical skills for reaching worldwide fame.

THE VALUE OF TECHNOLOGY

The subject matter of engineering ethics primarily concerns questions about the professional obligations engineers have by virtue of being engineers. To fully understand these obligations, it is helpful to ask some broader questions about the role of technology in society.

A few decades ago, many of the technologies we now take for granted, such as cell phones and the Internet, were beyond everyone's imagination. Many of us would probably say without much hesitation that these technologies have changed the world for the better. However, in the 1960s and '70s a series of engineering disasters triggered fundamental concerns about the value of engineering and technological development. It was widely realized that powerful technologies such as nuclear bombs and electronic surveillance systems can be used in morally problematic ways. Some argued that weapons of mass destruction and other military technologies help protect our freedom and independence, but others pointed out that the potential negative consequences could be so severe that it would be better to ban or abolish those technologies.

The debate over the pros and cons of *technology as such* is sometimes characterized as a debate between *technological pessimists* and *optimists.* Technological pessimists question the value of technological progress. A pessimist may, for instance, argue that we are no better off today than we were a hundred years ago, at least if we exclude medical technologies from the comparison. Although it is good to communicate with friends and relatives around the globe via the Internet, this technology also distracts us from valuable social interactions in real life.

Technological optimists point out that while it is true that some technological processes are hard to control and predict, and sometimes lead to unwanted consequences, the world would have been much worse without many of the technological innovations of the past century. Medical technologies save the lives of millions of people, and other technologies make it possible to travel fast and communicate with people far away. So even if it is sometimes appropriate to think critically about the negative impact of specific technologies, technological optimists believe we have no reason to question the value of technology *as such*.

The type of ethical questions technological pessimists and optimists discuss are very different from the moral questions LeMessurier sought to address during the Citicorp crisis. To clarify the difference between these issues, moral philosophers sometimes distinguish between *micro-* and *macroethical* issues.[7] This distinction is inspired by the analogous distinction in economics between micro- and macroeconomics. Microethical issues concern actions taken by single individuals, such as LeMessurier's actions in the Citicorp crisis. A macroethical issue is, in contrast, a moral problem that concerns large-scale societal issues, such as global warming.

In engineering ethics, discussions over the moral goodness or badness of specific technologies are examples of macroethical issues. If you, for instance, worry about a future in which autonomous drones equipped with artificial intelligence pose threats to lawful citizens, then your worry concerns a macroethical problem. Whether such autonomous drones will be developed does not depend on decisions taken by single engineers. What technologies will be developed in the future depend on complex social processes that no single individual can control.

The distinction between micro- and macroethics issues is useful for several reasons. It can, for instance, help us to understand discussions about moral responsibility. In some of the microethical issues discussed in this book, it is the engineer facing a tricky ethical choice who is *morally responsible* for the outcome. When LeMessurier decided to contact Citicorp to let them know that the tower had to be reinforced to withstand strong winds, it was LeMessurier himself who was morally responsible for doing the right thing. However, in discussions of macroethical issues, it is sometimes unclear who is morally responsible for a moral problem. Who is, for instance, morally responsible for making air travel environmentally sustainable? Whoever it is, it is not a single engineer. We will discuss the notion of moral responsibility in greater detail in chapter 12.

ENGINEERING ETHICS AND THE LAW

Questions about what engineers "may," "must," or "ought" to do have legal as well as ethical dimensions, but legal and ethical norms do not always overlap. Some actions are ethically wrong but legally permitted. You are, for instance, legally permitted to lie to your colleagues and to be rude to customers, but doing so is ethically wrong under nearly all circumstances according to every minimally plausible system of ethics.

The relation between ethical and legal norms is illustrated by the Venn diagram in Figure 1.4. Acts that are ethically wrong but legally permitted are represented in the left-most (light gray) part of the figure. Acts that are illegal *and* unethical are located in the overlapping (dark gray) area in the figure. In many cases our ethical verdicts coincide with the prescriptions of the law. In the United States (and many other countries), it is, for instance, illegal *and* unethical to discriminate against female engineers because of their gender. Another example is negligence. In engineering contexts, negligence can be defined as "a failure to exercise the care and skill that is ordinarily exercised by other members of the engineering profession in performing professional engineering services under similar circumstances."[8] Engineers who fail to meet this *standard of care*, that is, who are significantly less careful and skillful than their peers, violate the law and act unethically. If you, for instance, design a bridge, it would be illegal *and* unethical to propose a design that is significantly less safe than other, similar bridges designed by your peers.

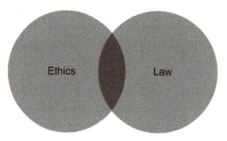

Figure 1.4

The leftmost part of the figure represents acts that are ethically wrong but legally permitted. The dark gray area represents acts that are illegal and unethical, and so on. The size of each area may not reflect the actual number of elements in each set.

The rightmost (light gray) part of the figure covers acts that are illegal but ethically permitted. Whether there really are any such acts is controversial. Socrates (469–399 BC) famously argued that it is *always* unethical to disobey the law. When he was sentenced to death for "corrupting the minds of the youth of Athens" by discussing philosophical problems, he stoically accepted this unfair ruling and drank the hemlock. A less extreme view is that we are ethically permitted to break laws that are clearly unjust, such as the Nazi laws against Jews during World War II. Another, less dramatic, example could be the case in which you could save the life of a dying baby by ignoring the speed limit on your way to the hospital.

The relation between ethics and the law can be further clarified by considering two prominent views about how moral claims are related to legal norms: *natural law theory* and *legal positivism*. According to natural law theory, morality determines what is, or should be, legally permissible and impermissible. On this view, legal rules are formal codifications of moral standards. It is, for instance, legally impermissible to use your employer's credit card for private expenses *because* this is morally wrong. There is a necessary connection between moral verdicts and their legal counterparts. Given that the moral aspects of a situation are what they are, the legal rules could not have been different. Putative counterexamples to this thesis, such as the observation that it is not illegal to lie to your colleagues or be rude to customers, can be dealt with by pointing out that legal norms must be simple and enforceable. It would be *impractical* to ban all types of immoral behavior.

Legal positivists reject the picture just sketched. They believe that law and morality are entirely distinct domains, meaning that we cannot infer *anything* about what is, or should be, legally permitted from claims about what is morally right or wrong. On this view, laws are morally neutral conventions. The fact that our laws tend to mirror our moral standards is a contingent truth. Every law could have been different. If we ask *why* something is illegal, the answer is not that it is so because the activity is morally wrong. The explanation has to be sought elsewhere: what makes a legal rule valid is, according to legal positivists, the fact that some political assembly has decided that the rule is valid and threatens to punish people who violate it.

Engineering ethicists seek to articulate moral rules, not legal ones; so in what follows, we will discuss legal aspects of engineering decisions only when doing so is helpful for understanding ethical issues. However, being legal or illegal alone never *makes* an act morally right or wrong, on either of the two views. To argue that it is morally impermissible for engineers to bribe business partners *because* bribery is illegal is a bad argument no matter whether one is a natural law theorist or legal positivist. According to natural law theorists, it is illegal to bribe people because it is morally impermissible. For the legal positivist, the two claims are conceptually unrelated.

ON ETHICS AND MORALITY

The word "ethics" stems from the ancient Greek word *ethikos* (ἠθικός), which means habit or custom. The word "morality" traces its origins to the Latin term *moralis*, which is a direct translation of *ethikos*. From an etymological perspective, the claim that an engineer's behavior is "ethically and morally wrong" is therefore no more informative than saying that it is "ethically and ethically wrong," or "morally and morally wrong." In this book, we shall respect the original meaning of the terms and use ethics and

CASE 1-2

Engineers Without Borders—Going Above and Beyond What Is Required

Moral commands sometimes go above and beyond legal rules. In April 2000, Dr. Bernard Amadei, a professor of civil engineering at the University of Colorado, was on a trip to Belize. He discovered that 950 Mayan Indians were living in the jungle without reliable access to clean drinking water. As a result of this, children could not attend school because they had to collect water from miles away. Although the community had the motivation and natural resources needed for solving the problem, Dr. Amadei realized they lacked the technical skills required for designing a lasting solution.

Back home in Colorado, Dr. Amadei formed a team of students and colleagues who built a functional prototype of a water system tailored to the needs of the Mayan Indians. Together with his team, he then went back to Belize to install a water system powered by a waterfall located a quarter of a mile from the village. This became the first project initiated and carried out by Engineers Without Borders USA (EWB USA).

Over the past two decades, EWB USA has grown considerably and currently has about 16,000 members, many of which are engineering students and academics. EWB International is an international association of sixty-three national EWB organizations, all of which strive to improve the quality of life for people in disadvantaged communities by designing and implementing sustainable engineering solutions. The international EWB network is inspired by, but not connected with, the French organization Doctors Without Borders (Médecins Sans Frontières).

Needless to say, no legal rules require engineers to devote their free time to voluntary work for Engineers Without Borders or other charity organizations. The decision to do so is entirely based on ethical considerations.

Discussion question: Do engineers have a professional obligation to volunteer for Engineers Without Borders and other similar organizations? Are there situations in which it would be wrong not to do so?

---

morality synonymously. This conforms to the dominant practice in applied ethics today. However, note that other authors reserve the term *ethics* for the *theoretical reflection* on moral rules that govern our everyday life. Nothing important hinges on this terminological issue as long as we make it clear how the words are used.

REVIEW QUESTIONS

1. Describe the main events of the Citicorp crisis. What was William LeMessurier's role? Did he, in your opinion, do anything morally wrong?
2. How is the practice of engineering regulated in the United States and Canada?
3. What do technological optimists and pessimists disagree about? Do you think it would be possible to settle this dispute by collecting empirical information about the consequences of using various technologies?
4. Explain the distinction between microethics and macroethics. Why might this distinction be (or not be) important?
5. Explain the relation between ethical and legal norms. Why do they not always coincide?
6. What is legal positivism? Do you think it is a plausible view?
7. A dictionary defines an engineer as someone who applies science and math to "make the properties of matter and the sources of energy in nature useful to

people." How should we understand the term "useful" in this definition? Are children's toys, computer games, weapons, or malfunctioning devices "useful" in the relevant sense?

REFERENCES AND FURTHER READINGS

Gayton, C. M., and R. C. Vaughn. 2004. *Legal Aspects of Engineering*. Dubuque, IA: Kendall Hunt.

Harris, C. E., Jr., M. S. Pritchard, M. J. Rabins, R. James, and E. Englehardt. 2014. *Engineering Ethics: Concepts and Cases*. Boston: Cengage Learning.

Herkert, J. R. 2001. "Future Directions in Engineering Ethics Research: Microethics, Macroethics and the Role of Professional Societies." *Science and Engineering Ethics* 7, no. 3: 403–414.

Morgenstern, J. 1995. "The Fifty-Nine-Story Crisis." *The New Yorker*, May 29, 1995, 45–53.

van de Poel, I., and L. Royakkers. 2011. *Ethics, Technology, and Engineering: An Introduction*. New York: John Wiley & Sons.

NOTES

1. National Society of Professional Engineers, *Code of Ethics for Engineers* (Alexandria, VA, 2018), 1. Available online: https://www.nspe.org/resources/ethics/code-ethics
2. The Institute of Electrical and Electronics Engineers, Inc., *IEEE Polices 2018* (New York, 2017), sect. 7.8. Available online: https://www.ieee.org/content/dam/ieee-org/ieee/web/org/about/whatis/ieee-policies.pdf
3. NSPE, *Code of Ethics*, 1.
4. Joe Morgenstern, "The Fifty-Nine Story Crisis," *The New Yorker*, May 29, 1995, 45–53.
5. This definition comes from *Merriam-Webster's Collegiate Dictionary*, 11th ed. (2014), Springfield, MA. Other dictionaries give similar but slightly different definitions.
6. This formulation of the industry exemption is from the Tennessee State Law. Other state laws formulate the industry exemption in slightly different ways, but the practical implications are roughly the same.
7. The distinction between micro- and macroethics has been carefully explored by Joseph R. Herkert. See, for instance, Herkert (2001).
8. J. Dal Pino, "Do You Know the Standard of Care?," Council for Advancement and Support of Education (CASE) White Paper (Washington, DC: American Council of Engineering Companies, 2014), 4.

Professional Codes of Ethics

Nearly every professional association for engineers has its own code of ethics. In this chapter, we discuss three of the most influential codes embraced by some of the largest engineering organizations in the United States: the NSPE, IEEE, and ACM.

NSPE stands for the *National Society for Professional Engineers.* It has about 31,000 members, and its mission is to address the professional concerns of licensed engineers (called Professional Engineers, or PEs for short) across all disciplines. IEEE stands for the *Institute of Electrical and Electronics Engineers.* As mentioned in chapter 1, the IEEE is the world's largest professional organization for engineers, with 395,000 members in nearly every country in the world. In addition to electrical and electronics engineers, thousands of computer scientists, software developers, and other information technology professionals are members of IEEE. ACM stands for *Association for Computing Machinery.* It has about 100,000 members, of whom about half reside in the United States.

Unsurprisingly, the codes adopted by the NSPE, IEEE, and ACM partly overlap. Some ethical concerns are unique to specific engineering disciplines, but many are generic. However, it would be overly simplistic to think that all ethical issues faced by engineers can be adequately resolved by looking up the "correct answer" in the relevant code. Many codes adopted by professional societies are better interpreted as ethical check lists. On such a reading, a code of ethics provides a point of departure for further analysis, but it does not by itself give the final answers to all ethical issues engineers may face in their careers. When confronted with some "easy" ethical problems, it may suffice to consult the relevant code; but in more complex situations, two or more principles may clash. In those cases, additional reflection and analysis is required, as explained in the last part of this chapter.

WHY ARE PROFESSIONAL CODES OF ETHICS IMPORTANT?

A code of professional ethics is a convention agreed on by a group of professionals. Why are such conventions worth your attention? And why are you morally obliged to comply with your organization's code of ethics?

A possible answer could be that you, when you decide to become a member of a professional organization, implicitly promise to abide by its code of ethics. To put it briefly, all members of the organization indirectly swear to behave in a certain way, meaning that every member ought to comply with the organization's code for the same reason as we, under ordinary circumstances, ought to keep our promises. The governing documents of IEEE stress that "Every IEEE member agrees to abide by the IEEE Constitution, Code of Ethics, Bylaws and Policies when joining."[1] The message is that if you voluntarily decide to enter this professional organization, you thereby indirectly promise the other members to follow its code of ethics.

The analogy with promises needs to be handled with care, however. Imagine, for instance, that you promise to do something very immoral: you promise to commit a crime. Never mind why you made this immoral promise; all that matters for the example is that you have promised to do something deeply immoral, which some people (such as criminal gang members) do from time to time. The fact that you have promised to commit a crime may give you *a* reason to keep your promise, but that reason is clearly outweighed by other, stronger reasons. All things considered, you ought not to do what you have promised. Therefore, it is not true that you always ought to fulfill your promises, under all circumstances, regardless of what you have promised. Whether we ought to keep our promises depends partly on *what* we have promised.[2]

Something similar applies to professional codes of ethics, although the moral requirements of a code are typically not controversial. The codes adopted by the NSPE, IEEE, and ACM have been drafted by highly competent professionals. There is no reason to think that any principle listed in those documents is immoral. However, during the apartheid era in South Africa, some local professional organizations developed deeply immoral, racist codes, which are no longer in force. This shows that the analogy with promises is valid. You clearly would have no obligation, all things considered, to comply with an unethical code. The mere fact that the members of an organization have agreed on a morally contentious code does not make the content of such a code right.

However, given that your professional organization's code does not demand you to do anything immoral, why should you then honor and promote the code if you already know how to tell right from wrong? This question is relevant because nearly no code comprises moral principles that new members of the organization are likely to be (entirely) unaware of before they enter the profession. Philosopher Michael Davis's answer is that engineers will be better off *as a group* if they all respect and promote the code of ethics adopted by their professional organizations, even if they already know how to tell right from wrong. Here is his argument:

> Without a professional code, an engineer could not object [to doing something unethical] *as an engineer*. An engineer could, of course, still object "personally" and

refuse to do the job. But if he did, he would risk being replaced by an engineer who would not object. An employer or client might rightly treat an engineer's personal qualms as a disability, much like a tendency to make errors. The engineer would be under tremendous pressure to keep "personal opinions" to himself and get on with the job.[3]

Davis's point is that engineers whose actions are governed by a code of ethics will find it easier to say no and thereby avoid temptations to make unethical choices. If *everyone* is required to behave in the same way, and all members of the organization know this to be the case, then engineers will find it less difficult to explain to clients and other stakeholders why they are unable to perform immoral actions. By acting as members of a group, each member will be more powerful than he or she would have been as a single individual. The personal cost of refusing to perform immoral but profitable actions decreases if one knows that everyone else will also refuse to do so.

Note that this argument works only as long as a *critical mass* of engineers remain loyal to the code. If too many members violate the code, then those who abide by it will be worse off than those who don't. Moreover, if an engineer can violate the code without being noticed or punished, he or she may sometimes have an *incentive* to do so. For reasons explained in the discussion of the prisoner's dilemma (at the end of chapter 5), it may hold true that although it is in the interest of every engineer to respect a professional code of ethics, rational individuals motivated by their self-interest may sometimes have compelling reasons to refrain from doing so.

THE NSPE CODE

The NSPE code begins with six *Fundamental Canons*, which are further specified in a number of *Rules of Practice* and *Professional Obligations*. The six Fundamental Canons follow. The entire document, with all of its Rules of Practice and Professional Obligations, is included in appendix A.

Fundamental Canons

Engineers, in the fulfillment of their professional duties, shall:

1. Hold paramount the safety, health, and welfare of the public.
2. Perform services only in areas of their competence.
3. Issue public statements only in an objective and truthful manner.
4. Act for each employer or client as faithful agents or trustees.
5. Avoid deceptive acts.
6. Conduct themselves honorably, responsibly, ethically, and lawfully so as to enhance the honor, reputation, and usefulness of the profession.

The moral principles expressed in a professional code of ethics typically fall within one of three categories. In an influential textbook on engineering ethics, Harris et al. introduce the terms *prohibitive, preventive,* and *aspirational* principles for describing the three types of principles.[4]

Prohibitive principles describe actions that are morally prohibited. For instance, the requirement that engineers shall "avoid deceptive acts" is a prohibitive principle.

If you violate that principle, you do something you are not morally permitted to do according to the NSPE code.

Preventive principles seek to prevent certain types of problems from arising. The second Fundamental Canon of the NSPE code (engineers shall "perform services only in areas of their competence") is a preventive principle. One of the Rules of Practice linked to this Fundamental Canon holds that "Engineers shall undertake assignments only when qualified by education or experience in the specific technical fields involved." At the beginning of the nineteenth century, engineers often overestimated their knowledge, causing numerous fatal accidents. By explicitly stating that engineers must be qualified "in the specific technical fields involved," many accidents can be avoided.

Finally, aspirational principles state goals that engineers should strive to achieve. An example of an aspirational principle is the sixth NSPE canon, according to which engineers shall "enhance the honor, reputation, and usefulness of the profession." It is not unethical to not enhance the honor, reputation, and usefulness of a profession, although it is desirable to do so if one can. No matter how much effort an individual engineer puts into realizing this goal, he or she will never reach a state in which the honor, reputation, and usefulness of the engineering profession is maximized. What makes these goals aspirational is the fact that they contain idealistic elements, which we will never actually achieve to the highest possible extent.

None of the principles listed in a professional code of ethics is legally binding. For instance, issuing public statements that are not "objective and truthful" is perfectly legal (but violates the third Fundamental Canon). Under most circumstances, engineers and others are legally permitted to lie and issue misleading public statements. A possible exception would be if you are called to testify in court. However, in the absence of written contracts or other legally binding agreements, engineers are almost always legally permitted to violate virtually every principle of their professional organization's code of ethics. However, and as noted in chapter 1, just because an action is legally permitted, it does not follow that you *ought* to perform that action. If you do something that violates your professional organization's code of ethics, then this is a *strong indication* that the action is unethical. Not every unethical action is illegal.

No code of ethics can cover all the unforeseeable moral problems that may confront an engineer, and nearly every code has multiple interpretations. A code of ethics is no full-fledged ethical theory. To illustrate the difficulty of applying a code to real-world cases, it is helpful to recall the Citicorp case discussed in chapter 1. William LeMessurier discovered in 1978, right before the start of the hurricane season, that the innovative skyscraper he had designed for Citibank in New York could topple in a moderate hurricane. LeMessurier quickly implemented a plan for strengthening the building by welding the bolted joints of the steel structure. For this he later received well-deserved praise. However, LeMessurier also misled the public about the true purpose of the nightly welding activities. In the press release jointly issued by Citibank and LeMessurier's office, it was stated that "there is no danger" to the public, which was an outright lie.

Let us analyze the Citicorp case though the lens of the NSPE code. Clearly, the press release violated the third Fundamental Canon of the NSPE code, which holds that engineers shall "issue public statements only in an objective and truthful manner." The full extent of this requirement is clarified by the following Rule of Practice (Appendix A,

"A.II. Rules of Practice," 3a): "Engineers shall be objective and truthful in professional reports, statements, or testimony. They shall include all relevant and pertinent information in such reports, statements, or testimony, which should bear the date indicating when it was current." Although LeMessurier was objective and truthful to his client, Citibank, he lied to the public and did not include "all relevant and pertinent information" in the press release.

Does this show that LeMessurier violated his professional obligations? No, that does not follow. LeMessurier's overarching aim was to comply with the first Fundamental Canon of the NSPE code, to "hold paramount the safety, health, and welfare of the public." It seems likely that the decision not to reveal the true extent of the structural problems helped to protect the public's safety, health, and welfare. Just like an airline captain may rightly choose not to disclose all details about a technical problem to the passengers to prevent panic, LeMessurier may have correctly concluded that full disclosure would have made it more difficult to complete the repairs on time. It is also possible that it was his client, Citibank, who explicitly asked LeMessurier to mislead the public. If so, LeMessurier could have defended himself by quoting the Rule of Practice (3a) listed in the NSPE code holding that "engineers shall not reveal facts, data, or information without the prior consent of the client or employer except as authorized or required by law or this code." LeMessurier was, arguably, not required by law or the NSPE code to reveal any information to the public.

To sum up, it seems that LeMessurier faced a conflict between the Rule of Practice holding that engineers shall "not reveal facts, data, or information without the prior consent of the client or employer" and the Fundamental Canon requiring that engineers must "issue public statements only in an objective and truthful manner." As explained at the end of this chapter, potential conflicts or inconsistencies in a code can be resolved in a number of different ways. There is no consensus on how engineers should behave if two or more principles in their professional code clash.

CASE 2-1

Gift/Complimentary Seminar Registration (BER 87-5)

Since its creation in 1954, the NSPE Board of Ethical Review (BER) offers ethical guidance to its members. The following is an anonymized case (one of approximately 500 cases) submitted to the Board for review over the past six decades:

The ABC Pipe Company is interested in becoming known within the engineering community and, in particular, to those engineers involved in the specification of pipe in construction. ABC would like to educate engineers about the various products available in the marketplace: the advantages and disadvantages of using one type of pipe over another. ABC sends an invitation to Engineer A, as well

as other engineers in a particular geographic area, announcing a one-day complimentary educational seminar to educate engineers on current technological advances in the selection and use of pipe in construction. ABC will host all refreshments, buffet luncheon during the seminar, and a cocktail reception immediately following. Engineer A agrees to attend.... Was it ethical for Engineer A to attend the one-day complimentary educational seminar hosted by the ABC Pipe Company?

The Board pointed out that the following Rule of Practice is applicable: "Engineers shall not solicit or accept financial or other valuable consideration,

(Continued)

directly or indirectly, from contractors, their agents, or other parties in connection with work for employers or clients for which they are responsible."[5]

If we were to interpret this rule strictly, it would follow that Engineer A should not be allowed to attend the one-day educational seminar because the buffet lunch and drinks were complimentary. However, in its ruling, the Board noted that this interpretation would be unreasonably strict:

> The Code unequivocally states that engineers must not accept gifts or other valuable consideration from a supplier in exchange for specifying its products.... However, in this case we are dealing with a material supplier who is introducing information about pipe products to engineers in the community and has chosen the form of an educational seminar as its vehicle. While ABC Pipe Company will seek to present its particular products in a favorable light and point out their many advantages over others', a complimentary invitation to such a seminar would not reach the level that would raise an ethical concern.... We note, however, that had Engineer A agreed to accept items of substantial value (e.g., travel expenses, multi-day program, resort location, etc.) our conclusion would have been quite different.

The Board's discussion indicates that the NSPE code is not an algorithm that can be applied mechanically. The code is a starting point for a deeper analysis. The rules of the code have to be interpreted by experienced professionals before a conclusion can be reached about a case.

Discussion question: Do you agree that the (monetary) value of the gift offered by ABC to Engineer A is morally important? If so, why?

THE IEEE AND ACM CODES

Unlike the NSPE code, the IEEE code does not distinguish between Fundamental Canons and Rules of Practice (see Appendix A). It consists of ten relatively straightforward principles:

> We, the members of the IEEE, in recognition of the importance of our technologies in affecting the quality of life throughout the world, and in accepting a personal obligation to our profession, its members and the communities we serve, do hereby commit ourselves to the highest ethical and professional conduct and agree:
>
> 1. to accept responsibility in making decisions consistent with the safety, health, and welfare of the public, and to disclose promptly factors that might endanger the public or the environment;
> 2. to avoid real or perceived conflicts of interest whenever possible, and to disclose them to affected parties when they do exist;
> 3. to be honest and realistic in stating claims or estimates based on available data;
> 4. to reject bribery in all its forms;
> 5. to improve the understanding of technology; its appropriate application, and potential consequences;
> 6. to maintain and improve our technical competence and to undertake technological tasks for others only if qualified by training or experience, or after full disclosure of pertinent limitations;
> 7. to seek, accept, and offer honest criticism of technical work, to acknowledge and correct errors, and to credit properly the contributions of others;
> 8. to treat fairly all persons and to not engage in acts of discrimination based on race, religion, gender, disability, age, national origin, sexual orientation, gender identity, or gender expression;

9. to avoid injuring others, their property, reputation, or employment by false or malicious action;

10. to assist colleagues and co-workers in their professional development and to support them in following this code of ethics.

The ACM code is more complex. Just like the NSPE code, it consists of a small set of foundational principles supplemented by a larger set of more specific rules. In what follows, we will focus on what the ACM refers to as its eight *General Moral Imperatives*. (The entire code is included in Appendix A.)

As an ACM member I will

1. Contribute to society and human well-being.
2. Avoid harm to others.
3. Be honest and trustworthy.
4. Be fair and take action not to discriminate.
5. Honor property rights including copyrights and patent.
6. Give proper credit for intellectual property.
7. Respect the privacy of others.
8. Honor confidentiality.

The principles adopted and advocated by NSPE, IEEE, and ACM express moral ideals many of us find highly plausible. However, as noted previously, the principles of a code may sometimes yield conflicting advice. The structure of such conflicts is usually the following: according to some principle X, you ought to perform some action A; but according to some other principle Y in the same code, you ought to not perform action A. How should such clashes between conflicting moral principles be dealt with and analyzed?

Case 2-2 (on the next page) is an example in which two of the principles in the IEEE code give conflicting advice. Unlike most other case studies in this book, this is a fictional case. The aim of Case 2-2 is to show that internal conflicts in the IEEE code are possible, not that they occur frequently.

Similar examples can be constructed for nearly all professional codes of ethics. It is, for instance, easy to imagine cases in which the second principle of the ACM code ("avoid harm to others") and the third principle ("be honest and trustworthy") clash. We just have to imagine that the *only way* to avoid harm to others is to *not* be honest and trustworthy. Here is an attempt to construct such a case: you know that the senior management of your company will do nothing to prevent a serious accident from occurring unless you exaggerate the magnitude of the threat somewhat. To do so would not be honest, but it would be necessary for avoiding harm to others.

CONTRIBUTORY REASONS AND MORAL DILEMMAS

How should engineers reason when two or more principles of a professional code of ethics give conflicting advice, or at least appear to do so? Here are four alternatives:

1. Reinterpret or reformulate the principles of the code.
2. Introduce a mechanism for resolving the conflict, for example, a hierarchical ranking of the principles.
3. Interpret the principles as *contributory* instead of *conclusive* moral reasons.
4. Accept the fact that it is sometimes impossible to comply with the code.

CASE 2-2

Cheating Diesel Emissions Tests

The details of the following case are based on the Volkswagen emissions scandal that surfaced in 2015, but some pieces of information have been modified with the aim of demonstrating that internal conflicts in the IEEE code are possible.

Imagine that you work for a US subsidiary of a large German automaker. Rumor has it that engineers responsible for the software controlling the company's diesel engines have manipulated the software. By using clever algorithms, the software detects if the car is being driven in regular traffic or is undergoing a test cycle of the type prescribed for the Environmental Protection Agency's (EPA's) emissions test. If the car senses that it is undergoing an emissions test, the software instructs the engine to develop very little power, which reduces the emissions. However, if the software senses that the car is being driven on a normal road, the software instructs the engine to work as normal, with emissions that are far above those allowed by the EPA.

When you contact your colleagues working in the software department about this, they assure you that everything is okay. They have not manipulated the software to mislead the EPA, they say. However, you find this hard to believe because it obvious that the emissions are higher than you would expect given the stellar performance in the official EPA emissions test. You conclude that your colleagues are probably lying. However, the only way to check this would be to perform some "false or malicious action." You could, for instance, mislead your colleagues to believe (falsely) that your manager has authorized you to review the source code. Or, if we wish to make the example more extreme, we could imagine that you bribe a junior employee in the software department to email you copy of the source code, and in return for this you invite him to a conference in a fancy hotel (see Figure 2.1).

Your method for verifying the truth of your hypothesis violates principle 9 of the IEEE code because you have, by "false or malicious action," obtained information that is very likely to injure "others, their property, reputation, or employment."

Figure 2.1
The software in this diesel engine senses whether the car is undergoing an emissions test. If so, it instructs the engine to develop very little power, which gives a misleadingly good test result. Is it morally wrong to obtain information about this form of cheating by some "false or malicious action" that is likely to injure "others, their property, reputation, or employment"?
Source: Dreamstime.

However, per principle 1 of the IEEE code, you did nothing wrong because you have merely fulfilled your obligation to "disclose promptly factors that might endanger the public or the environment."

The upshot is that no matter what you do, you will violate at least one principle of the IEEE code. The facts just happen to be such that you will either violate the first or the ninth principle of the IEEE code. There is no alternative action you could perform that would satisfy all principles of the code.

Discussion question: What would you do if no action available to you meets all principles of the IEEE code? What *should* you do?

The suggestion to reformulate or reinterpret the code is probably the option that first comes to mind. In the LeMessurier case, it could, for instance, be argued that the NSPE Rule of Practice holding that "engineers shall not reveal facts, data, or information without the prior consent of the client or employer except as authorized or required by law or this Code" should be *reinterpreted* to mean that it would have been permissible for LeMessurier to reveal all facts about the Citicorp building to the public, even if the client had instructed him not to do so. The reason for this could be the emphasis the NSPE code puts on holding paramount "the safety, health, and welfare of the public."

However, this maneuver is not likely to work in each and every possible case. Just consider the example discussed in the previous section. In situations in which the *only way* to avoid harm to others is to *not* be honest and trustworthy, no reasonable reinterpretation of the ACM code could restore its internal coherence and consistency. Another problem is that this strategy reduces the usefulness of the code. A code that cannot be read literally, because it needs to be interpreted or reformulated before it is applied, is less useful than one that doesn't require such maneuvers. Moreover, if the code can be interpreted in many ways, different members of the organization will likely prefer different interpretations.

A possible strategy for addressing these problems is to let a special committee (an ethics committee) decide how the code should be interpreted and applied. However, this usually takes time. An engineer who needs ethical advice *now* cannot ask a committee for advice about how to interpret the code.

The second alternative for resolving internal conflicts in a code is to introduce some mechanism that tells the engineer which principle to give priority to when two or more principles clash. It could be argued that the requirement in the NSPE code to "hold paramount" the safety, health, and welfare of the public should always carry more weight than all other principles. The remaining principles could be similarly ordered hierarchically, from the most important to the least important. Such a hierarchical approach could in theory resolve all potential conflicts built into a code.

Still, nothing indicates that the organizations advocating these ethical codes actually consider some principles superior to others. It seems *arbitrary* to insist that the principles should be ordered hierarchically. How do we tell which is most important: to be "honest and trustworthy" (principle 3 in the ACM code) or "avoid harm to others" (principle 2 in the ACM code)? It seems that a nuanced answer to this question should be sensitive to how *much* harm to others the engineer could avoid by not being honest and trustworthy. If you can save the lives of thousands of innocent citizens who are about to be killed in a terrorist attack by not being entirely honest when you inform your manager about the threat, then the requirement to avoid harm to others arguably outweighs the requirement to be honest and trustworthy. But if the amount of harm you can avoid is small, this may not be so.

Generally speaking, the problem with ordering conflicting moral principles hierarchically is that the relative weight of the conflicting principles may vary from case to case. Sometimes one principle is more important than another; but in other cases, some other principle carries more weight. This makes it difficult to determine a single, fixed, hierarchal order of conflicting moral principles.

The third method for addressing clashes between conflicting principles is to read the code as a list of *contributory* moral reasons instead of *conclusive* ones. A contributory moral reason speaks for or against an action but does not by itself entail any moral verdict about what one ought to do. For instance, if an engineer is lying, this could be a contributory reason for thinking that the action is wrong; but if the engineer lies to protect the health, safety, and welfare of the public, then this could be another contributory reason indicating that the action is not wrong. To tell whether the engineer's action is right or wrong *all things considered*, we have to balance all applicable contributory reasons against each other. Metaphorically, we can think of contributory reasons as "moral forces" that pull us in difference directions.

Conclusive moral reasons (sometimes called "overall" reasons) directly entail a moral verdict about the rightness or wrongness of the action. If, for instance, an engineer performs an action that violates someone's human rights, then that could be a conclusive (overall) reason that directly entails that the action is wrong.

The upside of thinking of the elements of a code as contributory reasons is that apparent conflicts can be resolved by balancing clashing principles against each other. What appears to be an irresolvable moral conflict may not always be so, all things considered. The downside is, however, that the code becomes less informative. It does not directly tell us what we ought to do. It is no easy task to balance conflicting contributory reasons, so engineers who seek guidance from the code may be not be able to figure out what to do by just reading the code.

The fourth and last method for addressing internal conflicts in a code is to accept the idea that it is sometimes *impossible* to comply with all principles. If, for instance, the only way you can avoid causing some large amount of harm is by not being entirely honest and trustworthy (as in Case 2-2), then every alternative open to you might be morally wrong. It is wrong to not act to avoid the harm, but the only way you can do so is by not being honest and trustworthy, which is also wrong. Acceding to the premises of the example, there is no third alternative action available to you that satisfies both principles.

Moral philosophers use the term "moral dilemma" as a technical term for referring to cases in which *all* alternatives are wrong. By definition, moral dilemmas are irresolvable: each option entails doing something morally impermissible. In everyday language, the term "moral dilemma" is often used in a broader and less strict sense. In such nontechnical contexts, a moral dilemma is a tricky situation in which it is difficult to know if some action is right or wrong. Everyday dilemmas are not moral black holes.

The best argument for thinking that genuine moral dilemmas exist is, perhaps, that we sometimes *feel* and *believe* that all options are wrong. Ethics is not like physics. It is simply false that conflicting "moral forces" can always be balanced against each other in a manner that makes at least one action come out as right. Moral principles or considerations are not like apples and pears on your kitchen scales: despite being very different, apples weigh at least as much as pears *or* pears weigh at least as much as apples. However, moral principles are multidimensional entities that resist binary comparisons. Although one alternative scores better with respect to one principle, and some other alternative scores better with respect to some other principle, thinkers who believe in moral dilemmas conclude that no action is, all things considered, entirely right

The Challenger Disaster

The Challenger disaster is one of the most well-known and widely discussed case studies in the field of engineering ethics. Although the accident occurred many years ago, we can still learn important lessons from it (see Figure 2.2).

On January 28, 1986, the Challenger space shuttle exploded shortly after take-off from Cape Canaveral, killing its crew of seven astronauts. Millions of people watched the launch live on TV. The direct cause of the crash was a leaking O-ring in a fuel tank, which could not cope with the unusually low temperature of about 26 °F (or −3 °C) at the day of the take-off. About six months before the crash, engineer Roger Boisjoly at Morton Thiokol, the company responsible for the fuel tank, realized that low ambient temperatures posed a potential hazard. Boisjoly wrote

a memo in which he warned that low temperatures at take-off could cause a leak in the O-ring: "The result would be a catastrophe of the highest order—loss of human life."[6] However, Boisjoly was not able to back up this claim with actual test results. His warning was based on extrapolations and educated guesses rather than hard facts.

At the prelaunch teleconference the night before take-off, Boisjoly reiterated his warning to NASA officials and colleagues at Morton Thiokol. Boisjoly was still unable to back up his concerns with data. No additional research into the O-rings' performance had been conducted since Boisjoly wrote the original memo. However, it was well known that small leaks, known as "blow-by," were common at higher temperatures. Boisjoly pointed out in the teleconference that it was reasonable to believe that there could

Figure 2.2
On January 28, 1986, the Challenger space shuttle exploded shortly after take-off from Cape Canaveral, killing its crew of seven astronauts. The official investigation showed that the cause of the disaster was a leaking O-ring in one of the fuel tanks.
Source: iStock by Getty Images.

(Continued)

be a correlation between the ambient temperature and the amount of leaked fuel. When pressed on this issue by the participants in the meeting, Boisjoly admitted that this was his professional opinion, not a proven fact. There were simply no hard facts available to him or anyone else about how the O-ring would perform at low temperatures.

After some discussion back and forth, the teleconference was temporarily suspended at the request of Morton Thiokol. The managers at Morton Thiokol knew that NASA would not launch the shuttle unless everyone agreed it was safe to do so. However, NASA was Morton Thiokol's largest and most important customer and NASA was under of lot of pressure. The managers at Morton Thiokol felt they needed a successful flight to secure an extension of their contract with NASA.

Just before the teleconference with NASA was about to resume and a decision had to be made, Morton Thiokol's Senior Vice-President Gerald Mason turned to Boisjoly's supervisor Robert Lund and said, "Take off your engineering hat and put on your management hat." The message was clear: because they could not prove it was unsafe

to launch, they had to look at this from a management perspective. Without a successful launch, the company's future could be at stake.

When the teleconference resumed, it was decided to go ahead with the launch. The shuttle exploded seventy-three seconds after take-off.

After the accident, Roger Boisjoly felt he had done everything he could to inform his supervisors about the problem with the O-ring. Despite this, he felt guilty for the tragic outcome. He was morally obliged to protect the safety of the astronauts, and he had tried his very best to do so, but he failed. Because he could not convince his supervisor that it was unsafe to launch, it was impossible for him to fulfill his obligation to the astronauts. A possible (but controversial) conclusion could thus be that Boisjoly was facing a moral dilemma: no course of action available to him in the fateful meeting was morally acceptable. According to this analysis, doing the best one can is not always sufficient.

Discussion question: Discuss Mason's comment to Lund, "Take off your engineering hat and put on your management hat." Under what circumstances, if any, should engineers reason like managers?

---

or morally permissible. Each alternative is, all things considered, somewhat wrong or impermissible from a moral point of view.

If you are faced with a genuine moral dilemma, no code of ethics will be able to resolve your predicament. Jean-Paul Sartre, a famous existentialist philosopher, claims that the only way to cope with such situations is to accept the fact that your dilemma is irresolvable. According to Sartre, "you are free, therefore choose, that is to say, invent. No rule of general morality can show you what you ought to do."[7]

PROPER ENGINEERING DECISIONS VERSUS PROPER MANAGEMENT DECISIONS

In hindsight, it is easy to say that it would have been good to listen to Boisjoly's unproven warning about the looming Challenger disaster. But was it *unethical* not to do so? And could the disaster have been avoided by paying more attention to professional codes of ethics?

To begin with, we should note that it would be unreasonable to maintain that safety concerns must *never* be ignored. For a somewhat extreme example, imagine that a drunk passenger on a cruise ship in the Caribbean warns the captain for the possibility of hitting an iceberg. Although it is true that icebergs can be dangerous, the captain has no reason to listen to the drunk passenger. The captain knows that there are no icebergs in the Caribbean, and the passenger has no relevant technical expertise. However, Boisjoly's warning was very different. Boisjoly was a technical expert on O-rings,

CASE 2-4

The Columbia Disaster

About seventeen years after the Challenger disaster, on February 1, 2003, space shuttle Columbia broke apart over Texas sixteen minutes before scheduled touchdown in Florida. All seven astronauts aboard were killed. During the launch two weeks earlier, a chunk of insulating foam had fallen off from the external fuel tank. The foam was soft and light; but because it hit the shuttle at a speed of 500 mph, the debris created a hole in the wing. This left a crucial part of Columbia with no protection against the heat generated by friction.

It was not uncommon for pieces of foam to fall off during the launch procedure. NASA called this "foam shedding." It had occurred in sixty-five of the seventy-nine launches for which high-resolution imagery was available; this was yet another example of normalization of deviance at NASA.

Rodney Rocha was NASA's chief engineer for the thermal protection system. While the shuttle was orbiting Earth, Rocha and his team studied images from the launch. They concluded that they would need additional images to assess how much damage, if any, had been caused by the foam. When Rocha suggested that NASA should ask the crew to conduct a spacewalk to inspect the shuttle, or ask the Department of Defense to use military satellites for taking high resolution images of the damaged area, his request was turned down. Rocha's supervisor Linda Ham told him that "even if we saw something, we couldn't do anything about it."[8] According to the investigation board's report, this statement was incorrect. Although it would have been difficult to repair the shuttle in space, the NASA team knew that another shuttle, Atlantis, was scheduled to launch within the next couple of weeks. According to analyses performed by NASA after the crash, it would have been possible to launch the Atlantis earlier than planned to rescue the astronauts aboard Columbia (see Figure 2.3).[9]

Figure 2.3
Space shuttle Columbia returns to Earth after a successful space mission. On February 1, 2003, it was destroyed during re-entry from space because of "foam shedding" during the launch procedure.
Source: NASA.

(Continued)

When Rocha realized that he would not receive the additional imagery he had requested, he drafted an email to his supervisor's supervisor. He never sent it to the intended recipient because he did not want to jump the chain of command, but he shared it with a colleague on the eighth day of the flight (January 22):

> In my humble technical opinion, this is the wrong (and bordering on irresponsible) answer from the SSP [Space Shuttle Program] and Orbiter not to request additional imaging help from any outside source. I must emphasize (again) that severe enough damage (3 or 4 multiple tiles knocked out down to the densification layer) combined with the heating and resulting damage to the underlying structure at the most critical location (viz., MLG [Main Landing Gear] door/wheels/tires/hydraulics or the X1191 spar cap) could present potentially grave hazards. The engineering team will admit it might not achieve definitive high confidence answers without additional images, but, without action to request help to clarify the damage visually, we will guarantee it will not. Can we talk to Frank Benz before Friday's MMT [Mission Management Team]? Remember the NASA safety posters everywhere around stating, "If it's not safe, say so"? Yes, it's that serious.[10]

Was the decision not to request additional images a proper engineering decision or a proper management decision? Clearly, because it required technical expertise and significantly affected the health, safety, and welfare of the astronauts, it was a proper engineering decision. It should have been made by engineers, not by managers. If NASA had followed Rocha's iterated recommendations to request additional images, it might have been possible to save the seven astronauts.

Discussion question: What can, and should, engineers do if they find out that the distinction between proper engineering decisions and proper management decisions is being violated?

and he expressed his concerns about the cold weather in his role *as a technical expert.* Boisjoly no doubt had the technical knowledge required for making an informed assessment of the risk.

This example suggests that what made Gerald Mason's decision to reject Boisjoly's unproven warning ethically problematic was the fact that it did not respect the distinction between what Harris et al. call *proper engineering decisions* and *proper management decisions.*[11] A proper engineering decision is a decision that (a) requires technical expertise; and (b) may significantly affect the health, safety, and welfare of others; or (c) has the potential to violate the standards of an engineering code of ethics in other ways. A proper engineering decision should be taken by engineers.

A proper management decision is a decision that affects the performance of the organization, but (a) does not require any technical expertise; and (b) does not significantly affect the health, safety, and welfare of others or has any potential to violate the standards of an engineering code of ethics in other ways.

When Gerald Mason told Boisjoly's supervisor Bob Lund to "take off your engineering hat and put on your management hat," Mason acted like a manager, even though it was clear that the decision to launch was a proper engineering decision. The purpose of the prelaunch meeting was to determine if it was safe to launch, but Mason used his management skills to reverse the burden of proof. Boisjoly was asked to prove that it was *not* safe to launch, which he was of course unable to do without conducting additional tests. If the distinction between proper engineering decisions and proper management decisions had been respected, the disaster could probably have been avoided.

Although the term "proper engineering decision" is not used in the NSPE code, the rationale behind this distinction is arguably influenced by the following NSPE Rule of Practice:

> If engineers' judgment is overruled under circumstances that endanger life or property, they shall notify their employer or client and such other authority as may be appropriate.

Boisjoly notified his employer about his concerns, but he did not contact anyone outside his organization. It could, perhaps, be argued that it was a mistake not to do so. (See the discussion of whistle-blowing in chapter 7.)

The official report on the Challenger accident was presented to President Reagan six months after the disaster. The Presidential Commission, led by William Rogers, pointed out that the problem with leaking fuel causing blow-by in the O-rings had been known and tolerated for several years. Sociologist Diane Vaughn coined the term *normalization of deviance* for describing the process in which a technical error is accepted as normal, even though the technological system is not working as it should. For another example of normalization of deviance, consider the "foam shedding" described in Case 2-4, which was the root cause of the Columbia disaster. This phenomenon was tolerated for years until it lead to the loss of seven astronauts.

REVIEW QUESTIONS

1. What reasons could engineers have for adopting a professional code of ethics?
2. Give some examples of prohibitive, preventive, and aspirational moral principles.
3. How do the IEEE and ACM codes differ from the NSPE code?
4. What is a moral dilemma (in a strict philosophical sense), and why could the concept of moral dilemmas be important for understanding the NSPE, IEEE, and ACM codes?
5. Summarize the main events of the Challenger disaster. What was the technical cause of the disaster and what ethical issues did it raise?
6. Explain the distinction between proper engineering decisions and proper management decisions.
7. Was the distinction between proper engineering decisions and proper management decisions respected when the Challenger was launched?
8. Did the decision to launch the Challenger violate any principles of the NSPE Code of Ethics?

REFERENCES AND FURTHER READINGS

Davis, M. Spring 1991. "Thinking Like an Engineer: The Place of a Code of Ethics in the Practice of a Profession." *Philosophy & Public Affairs* 20, no. 2: 150–167.

Gehman, H. W., Jr., J. L. Barry, D. W. Deal, J. N. Hallock, K. W. Hess, G. S. Hubbard, J. M. Logsdon, D. D. Osheroff, S. K. Ride, and R. E. Tetrault. 2003. Columbia Accident Investigation Board (CAIB). NASA: Washington, DC.

Gotterbarn, D. 1999. "Not All Codes Are Created Equal: The Software Engineering Code of Ethics: A Success Story." *Journal of Business Ethics* 22: 81–89.

Harris, C. E., Jr., M. S. Pritchard, M. J. Rabins, R. James, and E. Englehardt. 2014. *Engineering Ethics: Concepts and Cases*. Boston: Cengage Learning.

Institute of Electrical and Electronics Engineers, Inc. 2017. *IEEE Polices 2017*. New York, sect 7.

Marcus, R. B. 1980. "Moral Dilemmas and Consistency." *The Journal of Philosophy* 77: 121–136.

Martin, D. February 3, 2012. "Roger Boisjoly, 73, Dies; Warned of Shuttle Danger." *New York Times*.

McConnell, T. "Moral Dilemmas." *The Stanford Encyclopedia of Philosophy*. Edited by Edward N. Zalta. https://plato.stanford.edu/

National Society of Professional Engineers. 2007. *Code of Ethics for Engineers*. Alexandria, VA.

Rogers, W. P., et al. 1986. "Report of the Presidential Commission on the Space Shuttle Challenger Accident, Volume 1." Washington, DC: NASA.

Sartre, Jean-Paul. 1960. *Existentialism and Humanism*. London: Methuen, 30.

NOTES

1. The quote come from the section entitled "IEEE Governing Documents" on the IEEE webpage: www.ieee.org (last accessed October 8, 2018).

2. The ethics of promises is discussed in greater detail in chapter 6 in the discussion of Kantian ethics. What would Kant say about a case in which someone promises to do something immoral?

3. Michael Davis, "Thinking Like an Engineer: The Place of a Code of Ethics in the Practice of a Profession," *Philosophy & Public Affairs* 20, no. 2 (Spring 1991): 158.

4. Charles E. Harris Jr. et al., *Engineering Ethics: Concepts and Cases* (Boston: Cengage Learning, 2014), 14–18.

5. The rule quoted by BER is the formulation used by the NSPE in 1987. In later editions this rule has been split into two separate rules.

6. D. Martin, "Roger Boisjoly, 73, Dies; Warned of Shuttle Danger," *New York Times*, February 3, 2012.

7. Jean-Paul Sartre, *Existentialism and Humanism* (London: Methuen, 1960), 30.

8. Harold W. Gehman et al., "Columbia Accident Investigation Board (CAIB)" (Washington, DC: NASA, 2003), 152.

9. Ibid. 173.

10. Ibid. 157.

11. Harris et al., *Engineering Ethics*, 164–166.

CHAPTER 3

A Brief History of Engineering

Engineers have an immense impact on society. Few other professionals influence how we work, live, and interact with each other as much as people who design and develop new technologies. Before we discuss the moral implications of this, it is helpful to give a brief overview of the history of engineering. Have engineers always been shaping our lives to the same extent as they do today? Or is this a new phenomenon?

Historians seek to understand and explain change over time. By learning more about the history of engineering, we become better equipped to manage ethical issues related to the introduction of new technologies in society. However, there is no consensus in the literature on what causes technological change. Some influential historians, who shaped the discipline in the 1960s and '70s, subscribe to a view known as *technological determinism*. Inspired by the works of Karl Marx and others, they argue that technological innovations determine social developments. On this view, it would be a mistake to think that we can control technological transitions. Technology governs us; we do not govern technology. Consider, for instance, the nuclear bomb. If technological determinism is true, a wide range of political decisions made after World War II were ultimately triggered by the mere existence of the nuclear bomb. According to technological determinism, technology is the driving force in societal transitions and decision-making processes, and indirectly responsible for the development of new technologies.

Another group of historians, whom we will call *social constructivists*, believe that human decision makers shape technological innovation processes and give meaning to new technologies. On this view, human action drives societal transitions, as well as the development of new technologies. Consider, for instance, the telephone, invented in 1876. The invention of the telephone was not predetermined by some set of past events. On the contrary, the telephone was invented by Alexander Graham Bell as an aid for people hard of hearing. Bell could not foresee that it would be widely used by the general public for making long-distance phone calls. Social constructivists argue that this and other examples indicate that technology is socially constructed, in the sense that users often give technologies a different role or meaning than they were originally designed to have.

The *co-constructivist* stance is an intermediate position between technological determinism and social constructivism. Co-constructivists believe that technological innovations *together with* social processes shape social and technological transitions. Neither element is the sole cause. To explain change over time, we therefore need to pay attention to the technological as well as the societal dimension.

Because the co-constructivist position is the least radical alternative, and includes all the considerations emphasized by technological determinists and social constructivists, we will in what follows discuss the history of engineering from a co-constructivist perspective, meaning that we will highlight both technological and social processes for explaining change over time.

PREHISTORIC TECHNOLOGY

Prehistoric humans were hunter-gatherers. About 500,000 years ago, they began to use tools for a wide range of purposes. In graves in China and Europe, archeologists have found stone-tipped spears, cudgels, and other weapons, as well as stone tools for cutting and grinding wood. The archeological evidence also suggests that our ancestors knew how to make and transport fire. Prehistoric humans mostly lived in caves, and sometimes in primitive tents. They were nomads who followed migrating animal herds, meaning that they never settled in villages. Around 20,000 BC, the global population was about two million. (Today it is about eight billion; around two hundred cities in the world have a population exceeding two million.[1])

Although hunter-gatherers who made stone-tipped spears were by no means engineers, the fact that they used tools for hunting is worth emphasizing. Some other animals occasionally use tools, but no other species depends on tools as heavily and successfully as we do. Without tools, it would have been impossible to develop the technologies we now take for granted. In a sense, it is the fact that humans continuously develop new tools that make us humans. The stone-tipped spear was just the first step on a very long but remarkably fast journey.

ANCIENT ENGINEERING (5000 BC–500 AD)

Our ancestor's strategy for survival began to change around 12,000–10,000 BC. In the fertile river deltas in Mesopotamia in present-day Iraq, and along the Nile River, an agricultural revolution gained momentum. It was driven by new methods for soil tilling and by the domestication of dogs, sheep, goats, and cattle. This agricultural revolution made it possible to settle in villages because significant quantities of food could now be obtained in one and the same area. This led many humans to abandon previous nomadic practices.

Once settled down in villages, the settlers had stronger incentives than before to invest time and effort in improving their tools and agricultural technologies. If one knows that one will live in the same place for a long time, it becomes more attractive to think about how to ensure safe and continuous access to food in that location. As a result of this, irrigation methods were developed and improved, as were primitive wooden scratch plows that could loosen up the top two inches of the soil for planting. Clay was used for making bricks, and weaving techniques were also developed.

The wheel was invented around 4500–3500 BC in Mesopotamia, probably by a people called the Sumerians. The first wheels were made of wood and used in simple two-wheel carts, sometimes drawn by tame horses. This enabled the Sumerians to transport food and other goods over longer distances. This, in turn, made it possible to make the settlements larger. The Sumerians, who were relatively well-organized, built the first cities with temples and other large ceremonial structures. The Sumerians also discovered how to cast and mold bronze products, which enabled them to develop more efficient tools. In addition, they developed one of the first written languages.

The Egyptian civilization emerged approximately 3500 BC along the banks of the Nile River. The favorable agricultural conditions made it possible to produce large quantities of food with relatively little effort. The Egyptian society was well-organized, with clear (but nondemocratic) hierarchical structures. Egyptian rulers used the excessive supply of human labor for building cities and, famously, mausoleums in the form of gigantic pyramids.

All in all, about 80 pyramids were built. The largest is the Cheops pyramid in Giza near Cairo, completed around 2560 BC after about twenty years of construction. Its base measures 756 ft. (230 m), and it is 481 ft. tall (146 m). For nearly 3,800 years, it was the tallest man-made structure in the world until Old Saint Paul's cathedral in London was completed c. 1300 AD. The first building in the United States to exceed the height of the Cheops pyramid was the Philadelphia City Hall, completed in 1901.

The pyramids were built by thousands of slaves using ropes, sledges, rolls, and bronze chisels, but no wheels or draft animals. A man called Imhotep, who designed some of the earlier pyramids, is often credited with being the world's first engineer. Imhotep was also a priest and government minister.

Because the economic conditions for the Sumerians and Egyptians were different, it was attractive for the two civilizations to open trade routes. The Egyptians had plenty of wheat and papyrus, and they were skilled shipbuilders. The Sumerians produced large quantities of bronze and flax (which was useful for making textile). However, rather than trading directly with each other, a third group of people, the Phoenicians, acted as agents. The Phoenicians learned from the Egyptians how to build ships, and they also figured out how to navigate at night by looking at the stars. The Phoenicians eventually opened trade routes to several cities around the Mediterranean, which helped to quickly spread technical know-how from city state to city state.

Around 2500 BC, the population around the Indus River in today's India and Pakistan also began to grow. Just as for the Sumerians, the agricultural revolution made it possible to feed a larger population living in an urban area. Somewhat later, around 2000 BC, a similar agricultural revolution took place along the Yellow River in China, which also led to important technological innovations. The Indian and Chinese civilizations remained isolated from each other, as well as the rest of the world, so it took thousands of years before the Chinese inventions of ink, porcelain, and gunpowder reached the West.

The Sumerians never developed weapons or other technologies for warfare. This was a fatal mistake. Around 2000 BC, the Sumerians were defeated by the Akkadians, who used bows and arrows for conquering the Sumerian's land. The Egyptians made a similar tactical mistake. Around 1700 BC, they were invaded by the Hyksos, who used horse-drawn chariots and composite bows. The Egyptian technology of warfare (mostly spears and clubs) was outdated.

Around this time, numerous civilizations arose and prospered around the Mediterranean and the Middle East. The Greek city states were particularly important. The Greeks controlled a powerful navy, which could protect the city states against foreign invasions and keep trade routes open. During this period, iron replaced bronze as the metal of choice in tools and weapons.

While Egyptian engineers used trial and error for improving their technical know-how, Greek engineers relied on math and science. Pythagoras (c. 570–495 BC) is believed to be the first to *prove* that in every right-angled triangle, the square of the longest side is equal to the sum of the squares of the two shorter sides. Egyptians and others were familiar with this mathematical fact, but did not prove it. In *Elements*, a relatively short book written c. 300 BC, Euclid presented an axiomatic account of geometry. For more than two thousand years, all educated people in the West were expected to read and be familiar with Euclid's textbook. During this period, Greek thinkers also made good progress on some philosophical issues and invented a form of direct democracy (although women and slaves did not have voting rights). Plato founded the Academy in Athens, which was one of the world's first institutions of higher education.

The word "technology" stems from the Greek word *technê*, which can be translated as the craft or art to build or make things. By combining practical know-how with science, Greek engineers made several crucial breakthroughs. For instance, Archimedes (287–212 BC) discovered that the upward buoyant force of an object fully or partially immersed in a fluid is equal to the weight of the fluid displaced by the object. (Whether he took to the streets naked yelling "Eureka!" remains somewhat unclear.) Archimedes also constructed a large hollow screw for raising water, which could be used as a pump; and he built a water supply system that passed through a tunnel that was more than ½ mile long. Slave labor played a crucial role in all of these large-scale engineering projects.

Greek engineers built numerous temples, theaters, and other impressive structures, many of which still stand. One of the most famous and beautiful buildings is the Parthenon on the Acropolis in Athens. It was completed in 438 BC after just nine years of construction. A total of 22,000 metric tons of marble were transported from Mount Pantelakos, more than ten miles away, for building the temple. Compared to the pyramids, the Greek temples are more complex structures, which were significantly more difficult to build.

During this period, the Chinese civilization was relatively isolated from the rest of the world. To unify the country, but also to protect China from attacks by Mongolian nomads, Shi Huangti ("the First Emperor") ordered the construction of the Great Wall around 200 BC. Most of the wall that remains today was built much later, c. 1500 AD. The construction of the wall was to a large extent motivated by the political and social conditions in the region.

MEDIEVAL ENGINEERING (C. 500 AD–1400 AD)

The Roman Empire continued to expand until its sheer size made it difficult for a single person to govern it effectively from Rome. Around 300 AD, the empire was split in two halves. Rome continued to be the capital of the Western Roman Empire, while Byzantium (Constantinople, Istanbul) became the new capital of the Eastern Empire. The Eastern

CASE 3-1

Roman Engineering

Around 300 BC, the political and economic importance of the Greek civilization began to decline. The Romans gradually gained influence, mostly because their armies were larger and better equipped. The Romans made fewer scientific breakthroughs than the Greeks, but their engineers were remarkably successful. Over approximately 600 years, Roman engineers constructed a network of roads that stretched 60,000 miles and connected 4,000 towns and cities all over Europe. The city of Rome alone had a population of approximately 800,000–1,200,000 people.

Good roads were essential for supplying the capital with food and other goods. The roads were also used by merchants, tax collectors, and the army. Unlike roads built by previous civilizations, Roman roads could be used in all weather conditions and required little maintenance (see Figure 3.1). Normal highways were 12–15 ft. wide; main highways were up to 27 ft. wide. The layer of crushed stone, concrete, and sand used for constructing the highways was typically 4–5 ft. thick. Many of the Roman roads remained in use for more than a thousand years after the empire collapsed, with no or little maintenance. The poet Statius gave a detailed description of how the roads were constructed. Here is an excerpt:

> Now the first stage of the work was to dig ditches and to run a trench in the soil between them. Then this empty ditch was filled up with other materials viz. the foundation courses and a watertight layer of binder and a foundation was prepared to carry the pavement (*summum dorsum*). For the surface should not vibrate, otherwise the base is unreliable or the bed in which the stones are to be rammed is too loose.[2]

Roman engineers also built aqueducts and bridges that proved to be of exceptionally high quality, some of which are still in use. The main aqueduct that supplied Constantinople (present-day Istanbul, which was a Roman city at the time) with fresh water was more than 80 miles long.

Figure 3.1
Via Appia was a strategically important road connecting Rome with Brindisi in the south. Parts of Via Appia are still in good shape two thousand years after the road was constructed.
Source: iStock by Getty Images.

One of the key factors behind the Romans' many successful engineering projects was the invention of durable cement. Without cement, it would have been impossible to build the arches, domes, and vaults that are central to Roman architecture. One of the best examples is the Colosseum in Rome. This huge arena, which still stands at the center of Rome, seats 80,000 spectators and was completed 80 AD, after just eight years of construction. The stone blocks used for building the arena were transported more than twenty miles from Tivoli to Rome. The Colosseum was mainly used for gladiator games, but also for theater performances, public executions, and mock sea battles. As with many other large buildings from the ancient era, the Colosseum was built with slave labor, and more than 9,000 wild

(Continued)

animals were killed during the inaugural gladiator games. The Colosseum continued to be used until the sixth century. Its active service life was thus nearly 500 years. (The brand-new football stadium at Texas A&M University, the author's home institution, seats about 106,000 people. It will certainly not last for 500 years.)

Members of the Roman upper class enjoyed a luxurious lifestyle. Private homes were equipped with central heating. Many wealthy families owned a town house in Rome and a villa in the countryside. The natural meeting places in the bigger cities were the public baths, many of which were equipped with double glazing. Hot air circulated in tubes in the floor so it would remain warm.

None of these technological improvements occurred in a social or political vacuum. The Roman society was relatively stable and well-organized, which seems to be key to technological progress. Not everyone enjoyed a privileged lifestyle, and not everyone had the same opportunities; but the legal system was well developed and offered good protection to citizens. Business disputes could be settled in court, which made businessmen willing to invest in large construction projects. Many of the legal principles codified in Roman laws are still upheld today by modern courts. We cannot understand the technological achievements of the Romans without also understanding the legal, economic, and social conditions that made their large-scale engineering projects possible.

Discussion question: Can you think of any a modern building that you believe will last as long as the Colosseum in Rome? If not, why are modern buildings less durable? Is this something we should regret?

Empire continued to prosper from more than one thousand years until the Ottomans gained control of Constantinople in 1453. However, the Western Empire quickly began to decline. Historians disagree on why this happened, but some of the factors that may have played a role were increased military pressure from "barbarians" in the outskirts of the empire, increased corruption and inefficiency in the civil administration, increased social inequalities, and a string of weak emperors who were simply unfit for the job. The Western Roman Empire collapsed as a functioning state around 476–487 AD.

From an engineering perspective, the most important changes occurred in the Middle East. The teachings of the Prophet Muhammed spread quickly; and between approximately 700 AD and 1200 AD, Islamic technology was more advanced than that of medieval Europe. Scholars in the Middle East began to use Indian number symbols; and in addition to 1, 2, . . . , 9, they also introduced the number 0, which they noticed has some quite interesting properties. An obvious obstacle with the number system used by the Romans is that only experts can tell what MCXI times VII is. It takes less effort to see that 1,111 times 7 equals 7,777. Another important achievement was the introduction of standardized weights and measurements. (It is worth noting that engineering projects in the United States still fail from time to time because engineers forget to convert gallons to liters and feet to meters.) Islamic scholars studied classic Greek philosophy, and the Middle East became the intellectual center of the world.

Gunpowder was invented in China around 900 AD and reached Europe and Western Asia about three centuries later via Mongolia. China did not have much contact with the rest of the world, but this technology was so important that it quickly changed world history. Armies equipped with the biggest and latest cannons easily defeated armies with smaller and older ones, or none at all. For instance, Sultan Mehmet II conquered Constantinople (Byzantium, Istanbul) in April 1453 by using sixty-eight cannons that were 26 ft. long, each of which was equipped with a 1,200 lb.

stone cannon ball operated by a crew of 200 men. This marked the end of the Eastern Roman Empire.

In Europe, the iron plow was introduced around 600 AD. Unlike wooden plows, it cut deeper into the soil (up to 1 ft.). This triggered an agricultural revolution in Northern Europe. Previously it had not been possible to cut sufficiently deep into the soil in those heavily forested parts of Europe. Other important innovations were the iron horseshoe and the horse collar, which made it possible to use horses for a much wider range of purposes. Farmers also began to use blowing air (windmills) as a reliable power source for grinding harvests, spinning wool, and pumping water. The earliest references to windmills are from Arabic texts written before 951 AD. In Europe, several documents dated around 1180 AD mention windmills in Normandy and Yorkshire. We know with certainty that hundreds of windmills were built in Europe in the next one hundred years.

Flowing water (waterwheels) was another important power source. Although waterwheels had been used already by the Romans, this technology became increasingly important in medieval Europe. There were three types of waterwheels: vertical undershot wheels, vertical overshot wheels, and horizontal wheels. The simplest ones were vertical undershot wheels in which the water hit the paddles from beneath. The efficiency of these undershot wheels was low, approximately 22 percent, because most of the energy was lost to turbulence and drag. They were popular because they were simple to build and maintain. Vertical overshot wheels, in which the water hit the paddles from above, were more efficient (up to 66 percent) but also more difficult to build and maintain. The efficiency of horizontal wheels is estimated to have been about 40 percent (see Figures 3.2 and 3.3).

Figure 3.2
The photo shows a medieval vertical overshot wheel. Overshot wheels are more efficient than vertical undershot wheels and horizontal wheels (see Figure 3.3).   Source: iStock by Getty Images.

Figure 3.3
A horizontal waterwheel. Source: iStock by Getty Images.

During this period shipbuilding improved significantly. A new type of freight ship, the cog, first appeared in the tenth century. The Vikings also built other types of remarkably light and fast ships that were deployed for trips to North America in the West as well as Constantinople in the East.

Several impressive construction projects were undertaken during the medieval era. The best examples are the great cathedrals that still stand in many cities in Europe and England. Cathedrals were designed by skilled craftsmen who were in high demand. It was not unusual for the construction process to last for more than a century. Although the master masons who supervised the building process had no formal training (the newly founded universities in Bologna, Paris, and Oxford taught theology, medicine, law, and philosophy), we can think of these individuals as engineers. Many aspects of the construction process were very complex and required impressive amounts of expertise. However, unlike contemporary engineering, there were no theories that helped the master masons to predict what would work and what would not. Many cathedrals were built by trial and error. It happened from time to time that parts of the building collapsed before it could be completed.

EARLY MODERN ENGINEERING (c. 1400–1700)

During the early modern era, Europe became the most prosperous and fastest developing region of the world. Good ports and agricultural conditions led to increased trade. Combined with increased respect for private ownership, and a functioning legal system, a new class of wealthy private families gained influence. In cities such as

Venice, Dubrovnik, Milan, and Florence, successful businessmen built private palaces and sponsored artists and scientists. For instance, the Medici family in Florence (who owned the largest bank in Europe) supported Michelangelo as well as Leonardo Da Vinci. The latter was both a painter and engineer. He produced drawings for war machines, helicopters, pumps, and even a 650 ft. long bridge across the Golden Horn in Constantinople (Istanbul). Most of his designs were based on sound engineering principles; but because they were perceived as very radical, few of Da Vinci's machines were ever built.

One of the most important technological innovations of this era was Johannes Gutenberg's invention of the printing press in 1439. Up to that point, all books had to be written by hand, which made them very expensive, or printed with woodblocks, which were difficult to mass produce and led to fairly poor results. By using easily mass-produced metal types, which could be moved around and reused for different pages, as well as a new type of oil-based ink, it suddenly became possible to mass produce books at a fraction of the cost of a handwritten manuscript. In less than fifty years, ten thousand printing presses in Europe printed more than twenty-five million books. Before books could be mass produced, less than 5 percent of the population was able to read; but in just two generations, literacy increased to over 50 percent. Just as the Internet recently changed the way we communicate with each other, the printing press made it possible for new religious, political, and scientific ideas to spread much faster than before. Scientists such as Copernicus, Kepler, Descartes, Newton, and Leibniz could disseminate their theories quickly to a large audience of peers, which benefited scientific progress.

The need for pepper and other spices led explorers to search for new alternatives to the traditional Silk Road to the Far East. Vasco da Gama reached India by sea in 1498, and Christopher Columbus landed in America by accident on a failed attempt to reach the Far East by sailing around the world. These and other long journeys over the oceans were risky but immensely profitable. Only a fraction of the ships that set off from Europe came back with spices and other valuable goods. To manage these risks, merchants in Genoa, Amsterdam, and other commercial centers began to buy and sell insurance contracts. This enabled the owner of a ship to spread the risk over a larger group of investors. In the long run, everyone benefited.

Although many European universities had already been operational for centuries at this time, there were no institutions that offered formal instruction on topics relevant to engineers. One of the first institutions to do so was a school set up in Leiden by Prince Maurice of the Netherlands in 1600. The new school was connected with the university founded twenty-five years earlier, but it was not an official part of it. The professors in the university looked down on the engineering instructors, who they thought were too concerned with practical issues. This criticism was probably unfair. Dr. Ludolph Van Ceulen, who taught mathematics to engineering students at Leiden, published a book in which he correctly calculated π to twenty decimal places, which was of course of no practical value. Van Ceulen was clearly interested in mathematics for its own sake.

Many of the technologies that had been invented or significantly improved during the medieval era continued to be modified and refined. Ocean-going cargo ships grew larger and faster; and in many countries in Europe, canals were built, many of which had a large number of technically sophisticated locks. Wind and water in combination with draft animals continued to be the main power sources. America was colonized, and many technologies from the Old World spread quickly to the New World.

THE INDUSTRIAL REVOLUTION (c. 1700–1900)

The first successful steam engine was constructed in 1712 by Thomas Newcomen in England. It was used for pumping water from the bottom of a coal mine in Staffordshire so the miners could dig deeper and extract more coal. The idea that steam could be used as a power source had been around for a long time, but Newcomen was the first to successfully commercialize this game-changing idea. In the 1700s, there were no patent laws, so other engineers quickly develop bigger and more efficient steam engines. This was the starting point for what we today call the Industrial Revolution.

By the mid-1700s, about 100 steam engines were operating in coal mines in England; and by the end of the century, another 1,500 steam engines were in operation world-wide. This new power source suddenly made it possible to mass produce goods that had previously been manufactured manually. The existing power sources at this time were inefficient and unreliable (wind and water). Most people lived in rural villages in which goods were manufactured and consumed locally. The steam engine changed all this. Within a few decades, it became possible to generate enough power for sustaining larger cities and for producing and transporting large quantities of goods.

No other nation benefited as much from the Industrial Revolution as England. Around 1800, about 2 percent of the world's population lived there, but 50 percent of the world's goods were produced in steam-powered English factories. London was the largest and wealthiest city in the world, but also the most polluted one.

The riches created by the Industrial Revolution were not distributed equally in society. The living conditions for factory workers were miserable and often no better than those of their rural ancestors. The factory owners were, on the other hand, often able to amass huge fortunes. This led to social tensions. Karl Marx and Friedrich Engels, the authors of the Communist Manifesto (1848), lived and worked in England in the 1850s and developed their critique of the capitalist system there. It is also worth keeping in mind that the Industrial Revolution helped sustain the slave trade. Textile factories in England were dependent on imported cotton produced by slaves in the southern parts of North America.

Another consequence of the Industrial Revolution was that it became possible to travel. The first railroads, with wagons pulled by horses, were built in coal mines in England around 1790. A decade later, horses were replaced by steam engines; and in 1825, the first public, long-distance, steam-powered railroad was opened in northeastern England. It consisted of 25 miles of railway that connected the cities of Stockton and Darlington with a nearby coal mine. Because much larger amounts of coal could now be transported, at a much lower cost, the price of coal dropped about 50 percent in two years in Stockton.

In the following decades, railroads were built all over Europe and America (see Figure 3.4). By 1850, there were 9,797 km of railroads in Great Britain, 5,856 km in Germany, and 2,915 km in France. The first transcontinental railroad in the United States opened in 1869.

The maritime sector also underwent a steam powered revolution. Robert Fulton built the first commercially successful steam-powered riverboat in 1807, which carried passengers on the Hudson River between New York City and Albany. The first steam-assisted sail ship crossed the Atlantic in 1818 in as little as twenty-eight days. Unsurprisingly,

Figure 3.4

Railroads revolutionized the transportation sector in the nineteenth century. The photo shows a saddle-tank tender engine built by the Hunslet Engine Company, Leeds, England, in 1893.   Source: iStock by Getty Images.

steam-powered ships quickly got larger, faster, and more comfortable. Steamships also made it possible to sustain an ever-growing international trade. Imported goods became available to large groups of people and made it possible to import raw materials from distant shores.

During the Industrial Revolution, the demand for properly educated engineers increased. Up to that point, the military had been the largest and most important employer for engineers, but this changed as the demand for engineers in the private sector increased. The oldest non-military technical university in the world is the Technical University in Prague founded in 1707. At the end of the eighteenth century, technical universities were founded in Berlin, Budapest, Paris, Istanbul, and Glasgow. Just like today, students with engineering degrees were often able to secure well-paid positions and lead relatively comfortable lives.

Around this time, the public also got more interested in how engineers could change the world and people's living conditions. Some of the most successful engineers benefited from this increased interest in technology and gained public fame. As noted in chapter 1, superstar inventors such as Alexander Graham Bell, Thomas Edison, John Ericsson, and Isambard Kingdom Brunel became household names around the world. We still remember them for their technical skills, and their ability to commercialize new technologies, more than a century after they passed away.

CASE 3-2

Our Need for Energy: Electricity

During the nineteenth century, electricity became an increasingly important source of power, which eventually replaced the steam engine. The Italian physicist Alessandro Volta constructed the first battery in 1800, and the first electric dynamo (generator) was invented by Michael Faraday in 1831. However, it took an additional fifty years before this technology began to impact people's lives. Joseph Swan in England and Thomas Edison in the United States invented the lightbulb independently of each other and then joined forces in 1880 to install three thousand lightbulbs in London powered with DC (direct current) electricity from a steam-powered dynamo. The voltage of the first systems was 100 V, and the maximum distance of transmission was about half a mile. This limited the use of electricity to hotels, large private mansions, office buildings, steamships, etcetera. Nikola Tesla invented the first AC (alternating current) generator in 1884 and the AC motor in in 1888. Because AC can be transmitted over longer distances at high voltages, these innovations made it possible to use electricity for powering suburban trains and long-distance railroads. The first large-scale AC power generator, built in 1896, was powered with water from the Niagara Falls.

The battle between AC and DC systems continued for several decades, even though AC was technologically superior for most applications. This is one of the first examples of a technological lock-in effect. Because many users invested in DC systems at an early stage, it took many years before the technological advantages of AC systems made users switch to AC. Another example is the *qwerty* keyboard on modern computers, which is still in use because so many users have learned to use it. The fact that better, non-qwerty keyboards have been around for many years has so far had little effect on our choice between alternative writing technologies.

The discovery of electricity also revolutionized long-distance communications. The first telegraph was patented in 1837, the first undersea telegraph cable connecting England and France was installed in 1851, and the first transatlantic cable was connected in 1866. Before that time, it took weeks to send a message across the Atlantic. It is hard to overestimate the importance of this. Businessmen, politicians, and generals suddenly became able to control vast empires with branches on several continents from a central location. Newspapers were able to report about events happening in distant locations within hours instead of weeks. The societal importance of the step from no electric communication to the telegraph was arguably greater than the step from the telegraph to the Internet.

Discussion question: What actions, if any, do you think engineers can and should take to prevent technological lock-in effects?

MODERN ENGINEERING (C. 1900–)

Some scholars believe that the greatest invention of the twentieth century was the automatic washing machine (see Figure 3.5) and not the airplane, computer, or atom bomb.[3] The argument for this unorthodox claim is simple: by automating labor-intensive tasks traditionally carried out by women, the washing machine and other household devices enabled millions of women to enter the labor market, which in turn changed society more than ever before. In 1900, the labor participating rate for women in the United States was about 20 percent. Except for a short dip during the Depression in the 1930s, this number continued to increase steadily to about 65 percent in 2000. This was to a large extent possible because of technological developments that enabled people to devote less time to household chores, many of which had traditionally been carried out by women. By working outside the home, millions of women could for the first time in history earn their own living without being dependent on men. This led to demands for increased representation in legislative bodies, which in turn triggered numerous well-known social developments, including the invention of the Pill.

To understand and explain *how* and *why* all these technological transitions occurred, it is helpful to recall the three perspectives on the history of technology mentioned at the outset of the chapter: technological determinism, social constructivism, and co-constructivism. Can any of these perspectives offer plausible explanations of the many technological transformations of the twentieth century?

Let us keep focusing on the washing machine. Why was it invented, and how did it change society? The first patent for a washing machine was issued in 1791. By the mid-1850s, large, steam-powered, commercial washing machines were available in the United States and Europe. Those machines did, however, have relatively little impact on society. Most households continued to wash their clothes manually at home.

As electricity became more widely available, numerous manufactures de-

Figure 3.5
Why was the automatic washing machine invented, and how did it change society? The machine in the photo is believed to be from the 1940s.
Source: iStock by Getty Images.

veloped small, electric washing machines for domestic use. Ads for electric washing machines appeared in US newspapers in 1904, but we do not know for sure who invented the electric washing machine. However, we do know that in 1928, as many as 913,000 washing machines were sold in the United States. By 1940, about 60 percent of the twenty-five million households with electricity in the United States had an electric washing machine. The first automatic washing machines, in which the different phases of the wash cycle were executed without any input from the user, were introduced after World War II.

In a classic paper entitled "Do Machines Make History?" from 1967, Heilbroner explains that technological determinism is the view that "there is a fixed sequence to technological development and therefore a necessitous path over which technologically developing societies must travel."[4] Another influential formulation is the following: "[I]n light of the past (and current) state of technological development, and the laws of nature, there is only one possible future course of social change."[5] If a technological determinist were to analyze the history of the washing machine, the upshot would thus be that the electric and fully automatic washing machine was an unavoidable event in a process that started with steam-powered industrial washing machines in the 1850s and ended a century later with the liberation of women in Western democracies.

Determinists admit that there is no easy way to test the empirical accuracy of technological determinism. However, Heilbroner offers three arguments in support of his hypothesis:

1. The simultaneity of invention
2. The absence of technological leaps
3. The predictability of technology

The first argument, the simultaneity of invention, stresses that many discoveries are made by several people working independently of each other at roughly the same point in time. For instance, Joseph Swan in England and Thomas Edison in the United States invented the lightbulb independently of each other in the 1870s. The same day(!) as Bell filed his patent application for his telephone, on the 14th of February, 1876, Elisha Gray filed a preliminary patent application, known as a patent caveat, for a similar device. More generally speaking, most inventions seem to be made by several people when the sociotechnical conditions required for that particular invention have been met. If neither Edison nor Swan had invented the lightbulb, then someone else would have done so a few years later. The need for electric lightbulbs, and the availability of the key elements of this technology, meant that it was almost certain that *someone* would have invented the lightbulb around the time it was actually invented.

The second argument, which focuses on the absence of technological leaps, holds that technological advances tend to be incremental, contrary to what many of us may think. We almost never observe big technological leaps. It is true that some technologies change our living conditions dramatically, such as the atom bomb, which was invented in the early 1940s. However, Heilbroner points out that "we do not find . . . attempts to extract power from the atom in the year 1700."[6] This is because in the 1700s, the scientific steps needed for building the atom bomb had not been taken. No matter how hard a very smart person had worked on an atom bomb in the 1700s, he or she would have been doomed to fail. When we develop new technologies, we need to take one step at a time. This suggests that the development follows a predetermined path, with little room for extensive influence by individual humans. It would, for instance, have been impossible to invent the automatic washing machine before electricity replaced steam as the most significant source of power.

Heilbroner's third argument for technological determinism goes as follows:

The development of technical progress has always seemed *intrinsically* predictable. This does not mean that we can lay down future timetables of technical discovery, nor does it rule out the possibility of surprises. Yet I venture to state that many scientists would be willing to make *general* predictions as to the nature of technological capability 25 or even 50 years ahead. This, too, suggests that technology follows a developmental sequence rather than arriving in a more chancy fashion.[7]

This seems to be a reasonable claim. As the first electric washing machines were invented in the early 1900s, it was easy to predict that next step would be automatic versions. Moreover, as the first cell phones were introduced in the 1990s, it was easy to predict that improved models with bigger screens would soon follow, which is also what happened. Another straightforward prediction seems to be that within the next twenty-five years, almost all of us will be driving around in self-driving electric cars that require no or little input from the "driver." (I wrote this text in 2018. If you read this after 2043, you probably know if my prediction was correct.)

However, although the arguments discussed previously seem to offer *some* support for technological determinism, numerous historians have identified several weaknesses of this position. As mentioned before, many technologies are simply not used in the ways their inventors predicted and intended that they would be used. Moreover, many past predictions of future technological developments have turned out to be false. In the 1950s, scientists predicted that there would be small nuclear power plants available for domestic use, which we today know was false. Similar predictions about future space journeys have also turned out to be largely (but not entirely) inaccurate. Social constructivists argue that the best explanation of this is that there is no "fixed sequence" of technological development that follow a "necessitous path." On the contrary, individual as well as socioeconomic, cultural, ideological, and political dictions makers not only shape technological innovation processes, but also give meaning to new technologies. As human beings, we can freely decide what to strive for, try to invent, and invest our time and effort in. The automatic washing machine and other household devices became symbols for gender equality precisely because we gave this meaning to those technologies. We could have decided otherwise. Strictly speaking, it was not the automatic washing machine that caused the liberation of women but rather the meaning we gave to this technology. The physical aspects of the washing machine were not the driving force of history. What mattered was the beliefs, desires, and other cognitive attitudes people had toward those machines. In that sense, the washing machine is a social construction that drives a significant part of history during the twentieth century; and analogous remarks can be made about other technologies, such as airplanes or the atom bomb.

The co-constructivist stance is an intermediate position between technological determinism and social constructivism. To give a precise formulation of this approach is no trivial task, but here is an attempt. To start with, we note that technological determinists attribute the power to change history to a single primary force, namely, the technology itself. Social constructivists exclude the technology from the equation and instead attribute the power to change history to a multitude of social forces: socioeconomic, cultural, ideological, and political factors. To put it briefly, technological co-constructivists attribute the power to change history to all the factors identified by determinists and constructivists. Sometimes technological transitions follow a fixed sequence of steps, such as the transition from electric, low-voltage DC systems to high-voltage AC systems in the early 1900s. However, there are also examples of technological transitions that are mainly driven by, for instance, political factors, such as the current transition from traditional cars to electric ones. Each type of factor is important and sometimes plays a decisive role.

When formulated like this, the co-constructivist position is hard to reject. It is so accommodating that hardly anyone would disagree. This may, however, not be a blessing. If we cannot imagine any hypothetical empirical findings that would make us reject the co-constructivist position, it would have no empirical content and thus play no role in our understanding of history. Fortunately, the co-constructivists have a good response to this (Popperian) objection: if either determinists or constructivists could show that only one type of factor explains all technological transitions, then we should reject the co-constructivist position. As many of the examples discussed here indicate that this is not possible, it seems reasonable to conclude that technological as well as social factors are important drivers of many of the technological transitions observed throughout history.

REVIEW QUESTIONS

1. Discuss how knowledge about the history of engineering can be relevant for today's engineers.
2. Give a brief summary of the most important technological achievements by the Sumerians, the Egyptians, the Greeks, and the Romans.
3. Discuss what role engineers played for the rise and fall of the Roman Empire.
4. Give a brief summary of the most important technological achievements during medieval times.
5. Explain why the invention of gunpowder was so important, and compare it to the much earlier military innovations by the Sumerians and Egyptians.
6. Discuss the historical importance of Gutenberg's invention of the printing press in 1439. Was Gutenberg an engineer? Why or why not?
7. Explain, in your own words, the importance of the Industrial Revolution and the role engineers played in it.
8. Mention some of the most significant technical inventions during the Industrial Revolution.
9. Was it a coincidence that the steam engine was invented before the lightbulb, or could these and other events have occurred in a different order?
10. What is technological determinism? What are the best arguments for and against this view?
11. What are social constructivism (about technology) and co-constructivism claims *about*? That is, what question or questions do these theories seek to answer?

REFERENCES AND FURTHER READINGS

Harms, A. A., B. W. Baetz, and R. Volti. 2004. *Engineering in Time: The Systematics of Engineering History and Its Contemporary Context*. London: Imperial College Press.
Heilbroner, R. L. 1967. "Do Machines Make History?" *Technology and Culture* 8, no. 3: 335–345.
Hill, Donald. 2013. *A History of Engineering in Classical and Medieval Times*. London: Routledge.
Rosling, H. December 2010. "The Magic Washing Machine." TED Women, Washington DC.
Smith, M. R., and Marx, L. 1994. *Does Technology Drive History? The Dilemma of Technological Determinism*. Cambridge, MA: MIT Press.

NOTES

1. According to the United Nations, the total world population in 2017 was 7.55 billion people. See United Nations, Department of Economic and Social Affairs, Population Division, *World Population Prospects: The 2017 Revision, Key Findings and Advance Tables,* Working Paper No. ESA/P/WP/248 (2017), 1.
2. Donald Hill, *A History of Engineering in Classical and Medieval Times* (1984; repr., London: Routledge, 2013), 82.
3. See, e.g., H. Rosling, "The Magic Washing Machine" (TED Women, Washington, DC, December 2010).
4. Robert L. Heilbroner, "Do Machines Make History?," *Technology and Culture* 8, no. 3 (1967): 336.
5. B. Bimber, "Three Faces of Technological Determinism," in *Does Technology Drive History? The Dilemma of Technological Determinism*, eds. M. R. Smith and L. Marx (Cambridge, MA: MIT Press, 1994), 83.
6. Heilbroner, "Do Machines Make History?," 338.
7. Ibid.

PART

II

Ethical Theories and the Methods of Applied Ethics

CHAPTER 4

A Methodological Toolbox

There is little agreement on how ethical issues in engineering and other fields of applied ethics should be analyzed and managed. However, the absence of a single, universally accepted method does not entail that all methods are equally good, or that "anything goes." Some methods, views, and positions are clearly more coherent and nuanced than others. Reasonable people can disagree on whether an argument is convincing, but it does not follow that all possible ways of reasoning about ethical issues, including clearly flawed ones, are equally good.

In fact, most professional ethicists base their analyses of complex moral choice situations on a relatively small number of concepts, distinctions, and theories. Engineers who learn to speak the language of academic ethics are better equipped to formulate coherent and nuanced moral judgments. To improve our understanding of this terminology, it is therefore worth discussing some important distinctions, concepts, and theories in the field.

FACTS AND VALUES

The distinction between facts and values is central to nearly all discussions of applied ethics. People who disagree on some moral issue may do so because they disagree on what the relevant facts are, or because they do not accept the same value judgments. Imagine, for instance, that someone claims that the team of engineers who designed the Ford F-150 (see Figure 4.1), which is a large truck with a rather big engine, have a moral obligation to make it more fuel efficient. This is a moral claim, not a factual one. Scientific methods cannot determine whether we should accept it or not.

However, factual claims clearly play *some* role in this and other moral discussions. The following are examples of factual claims that seem to support the conclusion that

Figure 4.1

The Ford F-series is a best-selling vehicle in the United States, but it is less fuel efficient than many ordinary vehicles. Are the engineers who designed it morally responsible for this? Source: iStock by Getty Images.

the engineers who designed the Ford F-150 have a moral obligation to make it more fuel efficient:

1. The Ford F-150 is a large truck that is less fuel efficient than ordinary cars.
2. Large trucks are less environmentally sustainable than ordinary, more fuel-efficient cars.

The truth of premises 1 and 2 can be assessed by investigating the fuel efficiency of the Ford F-150 and its effects on the environment. However, whether premise 2 is true will depend partly on how we define the term "environmental sustainability." Reasonable people can disagree on the meaning of this term, as explained in chapter 16. Should, for instance, the notion of "environmental sustainability" be limited to effects on the natural environment, or should damage to ancient buildings (in, e.g., Athens and Rome) from car exhaust count as environmentally unsustainable effects? What this question shows is that in addition to factual and moral claims, we sometimes have to consider *conceptual* issues. This leaves us with the following three types of claims to consider in moral discussions:

1. Moral claims ("Engineers have a moral obligation to design environmentally sustainable technologies.")
2. Factual claims ("The Ford F-150 is a large truck that is less fuel efficient than ordinary cars.")
3. Conceptual claims ("The environmental sustainability of a technology does not depend on what harm it makes to manmade objects, such as ancient buildings.")

Factual claims have correct answers that depend on how the world is. The truth of a conceptual claim depends on the meaning of our concepts. Not all factual questions have answers that are known to us, and some may not even be *knowable*. Imagine, for instance, that someone asks whether the number of atoms in universe is odd or even. This question has an objectively correct answer, but it is virtually impossible for us to figure out what it is. We cannot count all atoms in the universe. Therefore, we will never be able to know whether the number of atoms is odd or even. The same problem arises if we ask how many tons of steel were produced in the world between January 1, 1800, and December 31, 1899. This question has an objectively correct answer, which does not depend on who is asking the question or for what purpose, but we will never be able to know the answer. The historical archives are, and will forever remain, incomplete.

Here are some examples of conceptual claims: "A bachelor is an unmarried man"; "A dangerous construction site is not safe"; and "All squares have five sides." The first two claims are true, but the third is false. Disagreement about conceptual claims can usually be resolved by considering the ordinary usage of a term or expression. It would, at least in some contexts, be odd to say that the damage made by exhaust fumes from cars to ancient buildings in Athens and Rome does not count as harm to the environment. However, technical terms such as "photonic crystal" and "covalent bond" usually get their meaning by stipulation. A covalent bond is a bond that involves the sharing of electron pairs between atoms *because* Irving Langmuir gave this meaning to the term in a scientific paper published in 1919.[1]

The Scottish philosopher David Hume (1711–1776) famously pointed out that many moral arguments begin with a set of nonmoral claims and then proceed to a moral claim without clarifying the relation between the moral and nonmoral claims (see Figure 4.2). Consider the following example:

Figure 4.2

David Hume (1711–1776) is one of the greatest philosophers of all time. He famously pointed out, among other things, that we cannot derive an "ought" from an "is." Source: Bequeathed by Mrs. Macdonald Hume to the National Gallery of Scotland.

1. The Ford F-150 is a large truck that is less fuel efficient than ordinary cars.
2. Large trucks are less environmentally sustainable than ordinary, more fuel-efficient cars.
3. Therefore, engineers who design large trucks, including the Ford F-150, have a moral obligation to make trucks more fuel efficient.

As noted, premises 1 and 2 are nonmoral claims. They do not by themselves entail any moral verdict. Whether we should accept the moral conclusion, 3, depends on what moral background assumptions we are willing to make. A possible view to take on 3 is that the only moral obligation car manufacturers have in a capitalist society is to produce vehicles that meet the environmental standards set by the regulatory authorities. On this view, it is up to the consumers to decide whether they prefer more fuel-efficient vehicles to less efficient ones, so long as the regulatory standards are met. Although some will certainly disagree with this moral standpoint, the example clearly shows that premises 1 and 2 do not *logically entail* 3. From a logical point of view, the acceptance of 1 and 2 does not force us to accept 3.

According to what is commonly known as *Hume's Law*, no moral claims can be derived from purely nonmoral premises. Let us apply Hume's Law to the Ford F-150 example. Clearly, the inference from premises 1 and 2 (of the list directly preceding this) to 3 is not valid. We need to *bridge the logical gap* between "is" and "ought" by adding an additional premise that links the factual claims about fuel consumption to the moral claim expressed in the conclusion. Here is an example of a possible bridge premise:

> (BP) Engineers working in the automobile industry have a moral obligation to improve the fuel efficiency of vehicles that are environmentally less sustainable than ordinary cars.

By adding (BP) to premises 1 and 2, the moral conclusion (3) is entailed by the premises; and so this maneuver addresses Hume's worry.

Of course, it remains an open question whether engineers actually *have* any moral obligation to improve the fuel efficiency of the vehicles they design. Hume's message was not that every bridge premise is morally acceptable. The point is just that by identifying the required bridge premise of a moral argument, the inference becomes logically valid. *If* you accept the premises, *then* you cannot rationally reject the conclusion. In many cases, this procedure for recasting a moral argument helps us to steer the discussion to what is the most central question of the debate: What moral obligations, if any, do engineers have for making vehicles more environmentally sustainable?

ARE MORAL CLAIMS OBJECTIVE, SUBJECTIVE, OR RELATIVE?

Factual claims are objectively true or false. If someone claims that "Maryam is an engineer," the truth of this claim does not depend on who says it or on that person's attitudes or beliefs. We can verify whether Maryam really is an engineer by contacting the institution that awarded her a degree.

But what about moral claims? Are they also true or false in an objective sense? Metaethics is the subfield of ethics that tries to answer questions about the nature and status of moral claims. Metaethical theories are thus theories *about* ethics, not theories about what someone ought to do or what actions are morally right or wrong. A wide range of metaethical positions have been proposed in the literature, and the terminology used by different scholars is sometimes confusing.[2] In this section, we will discuss four of the most prominent metaethical theories:

1. Ethical objectivism
2. Ethical nihilism
3. Ethical constructivism
4. Ethical relativism

Some authors consider ethical relativism to be a special version of ethical nihilism; but in what follows, we will treat it as a separate position.

Ethical Objectivism

Ethical objectivists believe that moral statements are true or false in just the same way as statements about tables and chairs, the weather, or the current unemployment rate are true or false. The truth of the claim that the unemployment rate is 5 percent depends on how many people are employed and not on what you and I say, or think, or feel. There are of course many ways of defining and measuring unemployment, and no definition or measurement process is entirely uncontroversial or politically neutral. However, once we have defined the term "unemployment rate" and clarified the meaning of this term, statements about the unemployment rate are objectively true or false.

Ethical objectivists also believe that the truth of ethical statements is *independent* of us, in the same sense that the truth of claims about the unemployment rate is, in a certain sense, independent of us. It is true that the unemployment rate depends on whether you and I work, so it is not entirely independent of us; but the unemployment rate does not depend on our *attitudes* or *wishes* or *feelings* about the unemployment rate. It is in this sense that ethical objectivists believe ethical judgments are valid independently of us.

So, if you are an ethical objectivist, what is right is right no matter what cultural traditions we have developed in our society. And what is right in country A is also right in country B, as long there are no morally relevant differences between country A and B.

Philosopher John Mackie raises two objections to ethical objectivism. The first is that ethical facts, if they were to exist, would be strange entities: they would be very different from all other types of facts we are familiar with. Where are ethical facts located, and what do they consist of? Mackie's point is that it is hard to imagine what kind of "thing" an ethical fact would be and how such facts could motivate us to act.

Mackie's second argument against objectivism is known as the *argument from disagreement*. It starts from the observation that there seems to be more disagreement on ethical issues than on ordinary factual issues. According to Mackie, the most plausible explanation of this observation is that there are no objective ethical facts. He writes that

> Radical differences between . . . moral judgments make it difficult to treat those judgments as apprehensions of objective truths [because] the actual variations in the moral codes are more readily explained by the hypothesis that they reflect ways of life than by the hypothesis that they express perceptions, most of them seriously inadequate and badly distorted, of objective values.[3]

Neither of these arguments is conclusive, but they have led a significant number of contemporary philosophers to believe that ethical objectivism is an untenable metaethical theory.

Ethical Nihilism

Ethical nihilism is the view that there are no objective ethical facts. Some nihilists are *expressivists*. If you say that "it is wrong to bribe government officials in foreign countries," then the expressivist would maintain that you are merely expressing an emotional attitude toward bribery. Your moral utterance has no truth value. It may be true that *you do not like* bribes, but your moral utterance only *expresses* this negative attitude. Because emotions are neither true nor false, expressivists believe that ethical judgments are neither true nor false.

Expressivists believe that moral disagreements are largely analogous to disagreements about music or food. Some people like green pesto a lot, but others don't. If you and I have different attitudes to green pesto, it would be wrong to say that one of us is right and the other is wrong. None of our attitudes is *true* or *false*, although it is of course true that we have the attitudes we have. In fact, there is not even any genuine disagreement here because we just have *different* emotional attitudes. We do not disagree on the truth value of any proposition.

Another type of ethical nihilism is the *error theory* proposed by Mackie. According to his proposal, all ethical judgments express genuine propositions about what ought and ought not to be the case; but all these judgments are false. It is, for instance, false that it is wrong for engineers to bribe clients and false that it is right to do so. No matter what ethical judgments we make, the error theorist would say it is a mistake to believe in the existence of objective ethical facts that make ethical judgments true. The "error" we make is that we speak and behave *as if* our ethical judgments express objective ethical truths, although no such truths exist.

Ethical Constructivism

Ethical constructivists believe that ethical statements have truth values that depend on socially constructed facts. A constructivist might claim that violating the rights contained in the United Nation's (UN's) Universal Declaration of Human Rights is wrong because most of the world's leaders have signed it; the signing made the moral claim true (see Figure 4.3).

If you are a constructivist, you could argue that the unemployment rate is, contrary to what we suggested previously, a social construction. It might be objectively true in an objectivist sense that a person is not working; but if we try to describe and measure this objective fact, we invariably *construct* a measure that essentially depends on our values and norms. There are many ways of measuring the unemployment rate, so what measure we use will depend on what (political) value judgments we wish to support. In that sense, the unemployment rate is a social construction.

Figure 4.3

Did the ratification of the UN Declaration of Human Rights in 1948 make it wrong to violate those rights, or would it have been wrong to do so even if the document had never been ratified? Source: iStock by Getty Images.

A problem with ethical constructivism is that this theory cannot easily explain why, for instance, Adolf Hitler's decision to kill more than six million innocent Jews during World War II was wrong. The Universal Declaration of Human Rights was ratified after World War II, so when Hitler committed his atrocities, it was *not* a socially constructed ethical fact that it was wrong to treat Jews in the way he did, at least not according to the UN declaration. If anything, it might have been morally right in this socially constructed sense because Hitler did in fact have a lot of support for his horrible views. This seems to be an unwelcome implication (of a somewhat naïve version) of ethical constructivism.

In response, the constructivist could point out that *other* social processes might have been in place before World War II that established the wrongness of Hitler's actions. Although many people supported Hitler's views (about 44 percent voted for the Nazis in the federal election in March 1933), the vast majority of Germans were Christians. A constructivist could argue that the *active practice* of a religion makes certain ethical claims true regardless of whether the religious claims as such are true.

Ethical Relativism

Ethical relativists believe that ethical claims are true or false only relative to some cultural tradition, religious conviction, or subjective opinion. Therefore, if Aysel claims that it is right to accept bribes and Bianca argues the opposite, both Aysel and Bianca could be making true ethical claims. What Aysel and Bianca claim might be true relative to their different cultures, religious convictions, or subjective opinions.

The difference between ethical relativism and ethical constructivism is that while constructivists believe that ethical truths can be objectively valid across different cultures or groups, in a sense that depends on us, relativists believe that the truth of ethical claims is *relative to* our culture or group membership, or something else.

A problem with many versions of ethical relativism is that this theory seems to make it impossible to criticize people who belong to other groups than our own for doing things that seem clearly wrong. Adolf Hitler is, again, a good example. What we want to say is that Hitler's atrocities were wrong, full stop. However, if we are ethical relativists, we instead must say that according to *our* standards, Hitler's actions were wrong—but according to the standards applicable in Germany during World War II, they were right. This is not a very attractive claim. It seems implausible to think that the wrongness of Hitler's atrocities could vary from one time and place to another. What Hitler did was ethically wrong, full stop, irrespective of what cultural or religious factors may have influenced his behavior.

Another problem for ethical relativists is that if we accept this theory, we could not even say that we *disagree* in any strong sense with Hitler about his attitude to Jews and other minorities. When we say that one should not murder innocent people because they belong to a minority group, we are just saying something that is true or valid relative to our own ethical standard. Hitler's claim that it was right to murder innocent people is also true, relative to his ethical standard. The two claims are therefore not in conflict with each other. It is a serious problem for ethical relativism that this theory makes genuine disagreement on ethical issues impossible.

APPLIED ETHICS AND ETHICAL THEORIES

Ethical theories seek to establish what features of the world *make* right acts right and wrong ones wrong. This is not the question applied ethicists seek to answer. In engineering ethics and other domains of applied ethics, the goal is to reach *warranted conclusions* about what to do in real-world cases given the limited and sometimes unreliable information available to the decision maker.

Ethical theories are often summarized in general moral principles: "Never treat others merely as a means to an end!" or "Only perform actions that lead to optimal consequences!" In chapters 5 and 6, we will discuss some ethical theories developed by Aristotle (384–322 BC), Immanuel Kant (1724–1804), and John Stuart Mill (1806–1873). Surprisingly, there is little agreement among applied ethicists on what role these and other theories should play for reaching warranted moral conclusions about real-world cases. In this section, we will discuss three alternative views.

According to the first proposal, every warranted moral conclusion about a particular case must be based on some ethical theory or general moral principle. This is the *theory-centered* view. The second proposal rejects the idea that warranted moral conclusions must be based on an ethical theory or moral principle. *Casuists* and *particularists* (we will discuss the difference between these positions shortly) believe that the tremendous variation across circumstances preclude the application of any universal moral principles on all cases. No general theory can tell us what to do—we must identify the morally relevant features of each specific case. Casuists and particularists believe that, strictly speaking, all general ethical theories are false.

The third proposal, which can be conceived as an intermediate position between the other two extreme positions, maintains that warranted practical conclusions can be based on some more *restricted* moral principles that are applicable to the ethical problems we are confronted with in engineering ethics, but perhaps in no other cases. Some examples of such *domain-specific* principles will be proposed at the end of the chapter.

The Theory-Centered Approach

According to the theory-centered approach, applied ethics is the application of some general ethical theory to a particular case. An example of such a general ethical theory is the utilitarian theory proposed by John Stuart Mill and others (see chapter 5.) Utilitarians believe, somewhat roughly put, that an act is morally right just in case it maximizes the total amount of well-being in the world, regardless of how this well-being is distributed in the world and regardless of the agent's intention.

Here is a schematic summary of how the utilitarian theory could be applied to William LeMessurier's decision to lie to the public about the Citicorp Center's wind-bracing system (see ch. 1).

1. Utilitarianism: An act is right if and only if it brings about at least as much well-being as every alternative act.
2. If William LeMessurier had revealed the truth to the public about the problem with the wind-bracing system, then that would not have maximized well-being. People living nearby would have felt fear and panic.
3. Therefore, it would not have been right of LeMessurier to reveal the truth to the public about the problem with the Citicorp Center's wind-bracing system.

A common worry about the theory-centered approach is that there is widespread and persistent disagreement about which ethical theory we should accept. Advocates of each of the ethical theories discussed in chapters 5 and 6 tend to insist that *their* theory is closest to the truth. However, in many cases, different ethical theories yield different recommendations. For the relativist, this may not be a great concern. If moral claims are valid only relative to the speaker's cultural or religious convictions, it is hardly problematic that there is no consensus on which theory we have the most reason to accept. Perhaps there is one optimal theory for each cultural or religious group. Expressivists may react in similar ways, although they would stress that there is, strictly speaking, no genuine *disagreement* here. Different theorists merely reflect different emotional attitudes; but because those theories are neither true nor false, it is misleading to describe this as disagreement. Some people prefer coffee and others tea, but there is not truth about what tastes best. According to the expressivist, the same applies to ethics.

However, for those who believe that moral claims are, in some sense, universally valid; and that, say, Hitler's actions during World War II were wrong irrespective of the agent's culture, religion, or personal convictions, it seems to be quite problematic that we do not know which theory we have the most reason to accept.

It has been proposed that we can deal with this *moral uncertainty* by assessing the probability that each ethical theory is correct and then maximize the expected moral value. To illustrate this idea, it is helpful to once again consider William LeMessurier's decision to lie to the public. He faced two options: tell the truth or lie. Let us suppose, for the sake of the argument, that LeMessurier assessed the probability that utilitarianism is correct to be 60 percent and the probability that Kant's theory of duty ethics is correct to be 40 percent. As explained in chapter 6, Kant believed in absolute rules according to which it is always wrong to lie, regardless of the consequences. From a utilitarian point of view, it was thus right for LeMessurier to lie at the same time as it was wrong from a Kantian point of view. Consider Table 4.1.

The idea that we can assign probabilities to ethical theories is not as exotic as one may think. We can take these probabilities to express judgments about the total evidence that speak for or against a theory, or the agent's degree of belief that the theory is correct. However, a more serious worry is that the assignment of moral values to the four possible outcomes seems to be arbitrary. How could we, even in principle, compare the value of performing an alternative that is wrong from a utilitarian point of view with that of performing an alternative that is right from a Kantian perspective? This appears to be an impossible, or even meaningless, comparison. It is thus unclear how a morally conscientious decision maker would react if confronted with the situation described in the table. Should the decision maker always do whatever has the highest *probability* of being morally right, or should she or he assign some kind of intertheoretically

Table 4.1 Moral uncertainty. The numbers represent the moral value of each outcome as assessed by the two theories.

	Utilitarianism (pr = .6)	Kantianism (pr = .4)
Tell the truth	Wrong	Right
Lie	Right	Wrong

comparable values to the four possible outcomes and maximize the expected value? If so, how could the morally conscientious decision maker determine those values?

Casuistry and Particularism

The word "casuistry" stems from the Latin term *casus*, which means "case." Casuists think that we can reach warranted moral conclusions about particular cases *without* accepting any ethical theory or moral principle. Casuists argue that the starting point of a moral analysis should be some set of judgments about cases we are already familiar with and know how to analyze. Such familiar cases are called *paradigm cases*. We know, for instance, that LeMessurier did the right thing when he decided to contact the owners of the Citicorp Center to inform them about the problem with the wind-bracing system (see ch. 1).

Imagine that we have been able to identify, say, ten paradigm cases. Casuists argue that any new case can be analyzed by comparing how *similar* it is to the paradigm cases we are already familiar with. The goal of this analysis is to treat like cases alike. For instance, if the new case is *fully similar* to a paradigm case, then the two cases should be treated in the same way. Moreover, if the new case is located somewhere "in between" two paradigm cases, then the analysis will depend on how similar the new case is to the two paradigm cases. The more similar to is to each paradigm, the greater influence should that case have on the analysis. If we come across new cases that differ "too much" from all known paradigm cases, we must rely on our practical judgment. Every situation is unique and has to be evaluated separately. There are no universal moral rules or principles that tell us what to do in all situations.

To illustrate the casuist's approach, imagine that you lie to a colleague on two separate occasions. The first time you lie because you believe it will benefit yourself. The second time you lie because you believe it will protect the well-being of the public. The two cases are similar in that they both involve instances of lying. However, the *reason* why you lied was different. It is, arguably, better to lie for altruistic reasons. So, when we compare how similar the cases are, we may conclude that the selfish lie is wrong and the altruistic one right.

Although many defenders of casuistry are Roman Catholics, the method itself need not be theological. A related approach is Jonathan Dancy's particularism, which also rejects moral principles. Dancy embraces a view called *reasons holism*: what counts as a reason for something in one case may count against it in another case. This applies to both theoretical and moral reasons. Suppose Alice sees an object that appears to her as red. Its red appearance provides her with a reason to believe the object is red. However, if Alice learns from Bob that a drug she has recently taken causes blue things to appear red and red things to appear blue, the polarity of her reason has changed. Now the fact that the object appears red does not count as a reason to think it is red but favors thinking it is blue. If the change of polarity is a general phenomenon of reasons, then it applies to moral reasons as much as theoretical reasons.

So, should we accept the particularist's position? It seems that particularists presuppose that some contributory reasons are assigned polarity-changing properties. In the example with red and blue objects, the second piece of information, that the observer has taken a drug that makes red things look blue and blue things look red, plays the role of such a reason. However, an alternative and more straightforward analysis of this example could be to say that the polarity of the first reason ("the object looks red") is the same before and after Alice is informed about the drug. On this alternative view, Alice's overall conclusion changes because her second reason is strong enough to outweigh the first reason. Therefore, it seems that we can explain everything that needs to be explained without postulating that whatever is a reason for doing something in one case can be a reason against doing the same thing in another case. And if this is true, the argument for particularism fails. If there exist some (moral) reasons that never change polarity, we can think of those "constant" moral reasons as moral principles.

Domain-Specific Principles

The third proposal mentioned previously is an intermediate position between the theory-centered approach and casuistry. According to this proposal, warranted moral conclusions about real-world cases can be reached by applying *domain-specific* moral principles.

Domain-specific principles are moral principles that apply to moral issues within a given domain but not to issues outside the domain. Domain-specific principles are less general than high-level ethical theories but not as specific as moral judgments about particular cases. (For this reason, domain-specific principles are sometimes called "mid-level" principles.) In medical ethics, the term *principlism* is sometimes used for describing this approach. In an influential textbook on biomedical ethics, Beauchamp and Childress propose four domain-specific principles for the biomedical domain: the autonomy principle, the non-maleficence principle, the principle of beneficence, and the justice principle.[4] If the four principles yield conflicting recommendations about a case, the agent has to balance the clashing principles against each other before a warranted conclusion about what to do can be reached.

Let us apply this proposal to engineering ethics. What would the most reasonable domain-specific principles for engineers be? Here are five plausible principles:

1. *The Cost-Benefit Principle (CBA)*
An engineering decision is morally right only if the net surplus of benefits over costs for all those affected is at least as large as that of every alternative.

2. *The Precautionary Principle (PP)*
An engineering decision is morally right only if reasonable precautionary measures are taken to safeguard against uncertain but nonnegligible threats.

3. *The Sustainability Principle (ST)*
An engineering decision is morally right only if it does not lead to any significant long-term depletion of natural, social, or economic resources.

4. *The Autonomy Principle (AUT)*
An engineering decision is morally right only if it does not reduce the independence, self-governance, or freedom of the people affected by it.

5. *The Fairness Principle (FP)*
An engineering decision is morally right only if it does not lead to unfair inequalities among the people affected by it.

These domain-specific principles are designed to apply to moral issues that arise within the engineering domain but not to other issues outside this domain. It would, for instance, make little sense to maintain that a federal judge should apply the cost-benefit principle for determining how many years a criminal should spend in prison.

As noted previously, a problem with domain-specific principles is that we sometimes have to balance conflicting principles against each other. How can we do this in a systematic and coherent manner? A possible solution is to think of domain-specific principles as objects in a geometric space. Consider Figure 4.4.

In Figure 4.4, each domain-specific principle covers a separate region of "moral space." Each principle is defined by a separate paradigm case. The notion of a paradigm case is roughly the same as that in casuistry. Once a paradigm case for each domain-specific principle has been identified, all other cases are compared to each other with respect to how similar they are to those cases from a moral point of view and then placed at the right location in moral space. The lines in the figure divide space into a number of regions such that each region consists of all points that are closer to its paradigm case than to any other paradigm case.[5] Within each region belonging to a paradigm case, the moral analysis is determined by the domain-specific principle corresponding to the paradigm case in question.

The moral principles sketched in Figure 4.4 are based on an empirical survey of the opinions of 583 engineering students at Texas A&M University. The participants were asked to compare ten cases with respect to their degree of (moral) similarity to other cases. The two-dimensional representation has been obtained through a multidimensional scaling of the similarities reported by the students. The interpretation of the two dimensions, freedom and uncertainty, was suggested afterward by the researcher (the author).

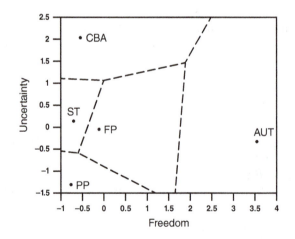

Figure 4.4

A two-dimensional representation of data elicited from 583 engineering students at Texas A&M University taking the course Ethics and Engineering. The horizontal dimension, freedom, varies from "no freedom at stake" (left) to "much freedom at stake" (right). The vertical dimension, uncertainty, varies from "much uncertainty" (down) to "no uncertainty" (up). CBA = Cost-Benefit Principle; PP = Precautionary Principle; ST = Sustainability Principle; FP = Fairness Principle; AUT = Autonomy Principle.   Source: M. Peterson, The Ethics of Technology: A Geometric Analysis of Five Moral Principles (New York: Oxford University Press, 2017).

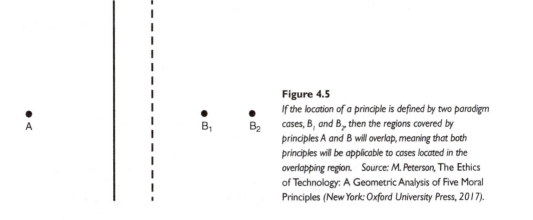

Figure 4.5
If the location of a principle is defined by two paradigm cases, B_1 and B_2, then the regions covered by principles A and B will overlap, meaning that both principles will be applicable to cases located in the overlapping region.   Source: M. Peterson, The Ethics of Technology: A Geometric Analysis of Five Moral Principles *(New York: Oxford University Press, 2017).*

The presentation of the geometric account of domain-specific principles outlined here is somewhat oversimplified.[6] For instance, in some cases it might be appropriate to represent moral principles as three- or n-dimensional regions. It is also worth stressing that if the location of a principle is defined by more than one paradigm case, then the regions covered by each principle will overlap, meaning that more than one principle will be applicable to the cases located in the "overlapping region," as illustrated in Figure 4.5.

REVIEW QUESTIONS

1. Explain the differences between moral claims, factual claims, and conceptual claims.
2. What is Hume's Law? Do you think it is valid? Can you think of any counterexamples?
3. What is metaethics? What relevance, if any, do metaethical theories have for discussions of engineering ethics?
4. What are the main features of the metaethical views known as moral realism, antirealism, and relativism?
5. What role, if any, could ethical theories play for resolving practical ethical issues faced by engineers?
6. How does moral uncertainty differ from other types of uncertainty?
7. What is casuistry and particularism?
8. What is a domain-specific moral principle, and how do such principles differ from moral theories such as Kantianism and utilitarianism?

REFERENCES AND FURTHER READINGS

Beauchamp, T. L., and J. F. Childress. 2001. *Principles of Biomedical Ethics*. New York: Oxford University Press.

Blackburn, S. 2002. *Being Good: A Short Introduction to Ethics*. Oxford, England: Oxford University Press.

Dancy, J. 2004. *Ethics Without Principles*. Oxford, England: Oxford University Press.

Hume, D. 1978. *A Treatise of Human Nature*. London: John Noon. First published 1739.

Lockhart, T. 2000. *Moral Uncertainty and Its Consequences*. New York: Oxford University Press.

MacIntyre, A. C. 1959. "Hume on 'Is' and 'Ought.'" *Philosophical Review* 68, no. 4: 451–468.

McNaughton, D. 1988. *Moral Vision: An Introduction to Ethics*. Oxford, England: Blackwell.

Peterson, M. 2017. *The Ethics of Technology: A Geometric Analysis of Five Moral Principles*. New York: Oxford University Press.

Railton, P. 1986. "Facts and Values." *Philosophical Topics* 14, no. 2: 5–31.

Railton, P. 1986. "Moral Realism." *Philosophical Review* 95, no 2: 163–207.

NOTES

1. Irving Langmuir, "The Arrangement of Electrons in Atoms and Molecules," *Journal of the American Chemical Society* 41, no. 6 (1919): 868–934.
2. Some authors prefer to distinguish between two positions: moral realism and antirealism. Moral realism is roughly equivalent to the position called ethical objectivism in this book. Moral antirealism is a broad terms for the three other positions discussed in this section.
3. J. Mackie, *Ethics: Inventing Right and Wrong* (Harmondsworth, UK: Penguin, 1977), 37.
4. Tom L. Beauchamp and James F. Childress, *Principles of Biomedical Ethics* (New York: Oxford University Press, 2001).
5. This type of diagram is known as a "Voronoi tessellation." Voronoi tessellations are use in a wide range of academic disciplines for representing similarity relations.
6. For a detailed presentation, see M. Peterson, *The Ethics of Technology: A Geometric Analysis of Five Moral Principles* (New York: Oxford University Press, 2017).

CHAPTER 5

Utilitarianism and Ethical Egoism

Ethical theories make general claims about what makes right acts right and wrong ones wrong. These claims are by their very nature explanatory. They tell us *why* some acts are wrong and others are right. Professional codes of ethics merely tell us *what* to do, but they seldom explain *why*. For example, we all agree that engineers ought to "avoid deceptive acts," as stated in the Fifth Fundamental Canon of the NSPE code; but *in virtue of what* are deceptive acts wrong? They are not wrong merely because the code says so. We may also ask if deceptive acts are wrong under *all* circumstances. Would it be wrong to perform a deceptive act if that were the *only* way you as an engineer could protect the safety, health, and welfare of the public? To answer these and other complex moral questions we have to consult ethical theories.

Ethical theories have a broader scope than professional codes of ethics. An ethical theory is designed to apply to *all* moral choice situations faced by engineers *and* everyone else. Ideally, an ethical theory should be to engineering ethics what math is to the engineering sciences: a solid foundation on which precise conclusions about real-world problems can be based.

Here is an analogy that is helpful for clarifying the difference between ethical theories and professional codes. Imagine that someone tells you that water boils at 100 °C (212 °F). This limited piece of information may lead you to falsely conclude that if you boil water at the top of Mount Everest, the boiling point would be 100 °C. This is not true because the boiling point of a liquid depends on the atmospheric pressure. To say that "water boils at 100 °C" is a simplification on a par with that we find in professional codes of ethics. Under *normal* circumstances, the statement is true. However, it is useful to know more about the laws of physics that govern phase transitions because that helps us identify the precise conditions under which the statement is *not* true. Ethical theories are designed to be universal and explanatory just as scientific theories. They are thus useful for answering questions that are overlooked by professional codes.

In this chapter, we will discuss two ethical theories that evaluate actions solely in terms of their consequences. The first one is utilitarianism, developed in its modern form by a group of 18th- and 19th-century English economists and philosophers. Cost-benefit analysis, discussed in chapter 9, is sometimes viewed as an application of utilitarianism to real-world cases. The second consequentialist theory is ethical egoism. Although this theory is rejected by many, the egoist's take on ethics is instructive. What reason do we have for thinking that we have moral obligations to others?

UTILITARIANISM

Utilitarianism holds that an act is right just in case no alternative act produces a greater sum total of well-being. It is the consequences, for everyone affected, that make the act right or wrong.

The first modern utilitarian was Jeremy Bentham. His book *Introduction to Principles of Morals and Legislation* (1789) is still widely read by utilitarians.[1] John Stuart Mill and Henry Sidgwick are two other utilitarian thinkers whose works continue to be important in the contemporary debate. Bentham defines utilitarianism as follows:

> By the principle of utility is meant that principle which approves or disapproves of every action whatsoever, according to the tendency it appears to have to augment or diminish the happiness of the party whose interest is in question.[2]

Bentham stresses that it is not merely the happiness or well-being for the agent who performs the act that matters. Bentham and other utilitarians are deeply committed to the idea that everyone's happiness or well-being matters equally much. If you face a choice between doing something that makes you yourself one unit happier or improving the situation equally much for a stranger you have never met before, utilitarians would argue that both options are equally valuable from a moral point of view.

It is often pointed out that utilitarianism sometimes conflicts with conventional morality and deeply rooted cultural norms. For instance, utilitarians have to admit that it is not always wrong to lie. If lying brings about the best overall consequences, then that is the right thing to do. Consider, for instance, a hypothetical case in which an engineer's decision to lie to a client turns out to prevent something very bad from happening: in a meeting at the Hilton in New Orleans, electrical engineer Josh persuades his client to spend $1 million extra on safety measures the client would otherwise have ignored. Josh's method for persuading his client is to knowingly exaggerate the probability that the building he is designing will flood during a hurricane. To knowingly give false information by exaggerating a hazard counts as a form of lying. Due to a series of unforeseen circumstances, the $1 million extra spent on safety measures later turns out to save the lives of dozens of people when the next hurricane hits New Orleans.

Conventional morality prohibits nearly all lying, but the utilitarian theory suggests a more nuanced picture. If Josh's lie to his client brought about better total consequences for all those affected, then this lie was not wrong. The fact that conventional morality condemns Josh's behavior is irrelevant. Utilitarians believe that the consequences of an act are the only relevant considerations for determining what is right. If the overall consequences of lying—that is, not just the consequences for the person who tells the lie but all the consequences for everyone affected by the act—are better than telling the truth, then it is morally right to lie. That said, there are of course numerous other situations in which the utilitarian theory gives less controversial (and perhaps more plausible) recommendations.

CASE 5-1

Prioritizing Design Improvements in Cars

In the mid-1990s, the US National Highway Traffic Safety Administration (NHTSA) considered whether various automobile design improvements should be mandatory for the manufacturers. It was estimated that the cost of implementing each design improvement, expressed in dollars per life-year saved, would have varied between $0 and $450,000 (see the list that follows). The benefit of each design improvement was the same: the prevention of one statistical death per year per dollar amount listed in the table. Granting a limited budget, how should these improvements be prioritized?

1. Install windshields with adhesive
   bonding $0
2. Mandatory seat belt use law $69
3. Mandatory seat belt use and
   child restraint law $98
4. Driver and passenger automatic
   (vs. manual) belts in cars $32,000
5. Driver airbag/manual lap belt
   (vs. manual lap/shoulder belt) in cars $42,000
6. Airbags (vs. manual lap belts) in cars $120,000
7. Dual master cylinder braking system
   in cars $450,000

Utilitarians are ultimately concerned with maximizing well-being and not dollars. Because money and well-being are different things, utilitarians have no direct reason to minimize the number of statistical deaths in traffic accidents. However, it is reasonable to assume that in this situation, utilitarians would conclude that the more statistical deaths we prevent (per dollar spent on traffic safety), the more well-being would be created in society. If so, utilitarians should rank the alternatives in the order just listed. Hence, installing windshields with adhesive bonding would have maximized well-being, followed by a mandatory seat belt use law.

Critics of utilitarianism would disagree with this ranking. For instance, although all fifty states in the United States now have seat belt laws, natural rights theories would argue that laws that make seat belts mandatory violate the individual's right to self-ownership. If you own yourself, then this entails a right to choose freely what risks to accept. On this view, the utilitarian ranking incorrectly omits the fact that a mandatory seat belt use law limits your freedom to decide if the risk of not using your seatbelt is worth taking. (Cf. the section on natural rights in chapter 6 and the discussion of cost-benefit analysis in chapter 9.)

Discussion question: Is it possible to assign monetary values to rights violations? Suppose, for instance, that you have a right to decide whether to use the seat belt in your car. What is the "cost" of a law that violates this right by making it mandatory to use the seat belt?

Modern philosophers have clarified the utilitarian theory by making a number of important distinctions. First, contemporary utilitarians distinguish *particular acts* from *act types*. The particular act of lying in the hurricane example turned out to be good. By lying to his client, Josh saved dozen's of people. It is reasonable to assume that this positive consequence is worth more, morally speaking, than the extra money Josh's client spent on safety measures. Therefore, the particular act Josh performed when he lied to his client in the meeting at the Hilton in New Orleans were good. However, many similar particular acts would have radically different consequences. If Josh were to lie to another client a few years later, under similar circumstances, the overall consequences could very well be bad because the second lie might not save anyone's life. If you are an act utilitarian, then you believe that an act's moral features depend on the properties of the *particular act* the agent performs and not on the features of the type of act of which it is an example.

An *act type* is defined by the central properties shared by all similar particular acts. The particular act of lying to a client in the meeting at the Hilton in New Orleans belongs to the act type "lying to clients." Viewed as an act type, lying to clients has bad consequences because most acts of this type have negative consequences. It is, perhaps, not surprising that nearly all classic and contemporary utilitarians think we should evaluate particular acts, not act types.

The next distinction that helps us to better understand the utilitarian theory is that between, on one hand, the *actual* consequences of a particular act and, on the other hand, the *foreseeable* or *expected* consequences. The foreseeable or expected consequences can be determined by calculating the probabilities of all possible consequences and multiplying these numbers by the value of each consequence. (See Case 5-2 *Climate Change and Human Well-being* for an explanation of how to do this.) It is reasonable to assume that the expected or foreseeable consequences of Josh's lie to his client in the meeting at the Hilton in New Orleans were bad because he had little reason to believe that the lie would have the consequences it actually had. However, the *actual* consequences of Josh's act were good because his decision to lie to his client turned out to save dozens of lives.

There is no consensus in the literature of whether it is the actual or expected consequences that matter from a moral point of view. A possible argument for thinking that it is the actual consequences that matter is the following: if we look back in time and are offered to alter the past(!), it seems that what we would have reason to alter is the actual consequences of an act and not the expected consequences. What we ultimately care about, and should care about, is how things *actually go*, not how we expect things to go. However, a reason for focusing on expected consequences could be that it might then be easier to gain knowledge about what is right and wrong. This is a reason that some (but not all) moral theorists consider to be important.

The third and final distinction utilitarians consider to be crucial is that between act utilitarianism and rule utilitarianism. Some thinkers are rule utilitarians. They believe that we ought to act per a set of rules that would lead to optimal consequences if they were to be accepted by an overwhelming majority of people in society. Rules such as "Don't lie!" and "Don't harm innocent people" are examples of rules that could be included in such a list. If an overwhelming majority in society were to accept those rules, that would arguably have optimal consequences. The rule utilitarian theory is attractive partly because it tallies well with our everyday moral intuitions. Many of us believe it is wrong to lie, or harm innocent people. Unlike other versions of utilitarianism, the rule utilitarian theory can explain why: the consequences are, on average, suboptimal. Another reason for accepting rule utilitarianism could be that it is easier to learn to act in accordance with a small set of rules than to evaluate every new act on an individual basis. Once we have established what the optimal rules are, it will be relatively easy to figure out if an act is right or wrong.

Act utilitarians question the moral relevance of what would happen if we were to act in accordance with some set of carefully selected rules. Act utilitarians believe that the right-making features of an act are the consequences of that particular act, not the hypothetical effects of rules that may or may not be followed by others. Note, however, that act utilitarians may concede that what has best consequences on a particular

CASE 5-2

Climate Change and Human Well-being

According to the Fifth International Panel on Climate Change (IPCC)

> The atmospheric concentrations of carbon dioxide, methane, and nitrous oxide have increased to levels unprecedented in at least the last 800,000 years. Carbon dioxide concentrations have increased by 40% since pre-industrial times, primarily from fossil fuel emissions and secondarily from net land use change emissions. [. . .] Human influence on the climate system is clear. This is evident from the increasing greenhouse gas concentrations in the atmosphere, positive radiative forcing, observed warming, and understanding of the climate system. [. . .] It is extremely likely that human influence has been the dominant cause of the observed warming since the mid-twentieth century.

How would utilitarians analyze alternative strategies for mitigating climate change? Utilitarians see nothing of moral concern in the increasing average global temperature in itself. We have no moral obligation to the Earth as such. However, utilitarians worry about how climate change affects the well-being of present and future human beings and other sentient beings. To illustrate, imagine that continuing to emit high levels of greenhouse gases would be worth 10 units of well-being for the present generation, while all future generations would be deprived of 1,000 units in total. (These numbers

Table 5.1 A hypothetical utilitarian analysis of climate change. The numbers refer to the aggregated well-being for present and future generations

	Our generation	Future generations	Sum total
High emissions	+10	−1,000	−990
Low emissions	−20	+500	+480

are fictive numbers chosen for illustrative purposes.) The net effect would be −990 units. Drastically lowering the emissions would cost the current generation −20 units, while future generations would gain 500 units. The total consequences for everyone would thus be +480 units. See Table 5.1. According to the utilitarian theory, it would therefore be wrong not to lower emissions.

In this example, the benefits for future generations clearly outweigh the cost for the current generation. Utilitarians do not permit us to give more weight to our own well-being. Everyone's well-being counts equally, no matter whose well-being it is.

Discussion question: Utilitarians are concerned with how climate change affects the well-being of present and future sentient beings. The fact that the ice is melting in the arctic is in itself of no direct concern to utilitarians. Do we have an obligation to preserve nature as it is, or is it permissible to perform actions that change the climate as long as there is no net harm to sentient beings?

---

occasion might be to follow a rule: if you act in accordance with the rule "Don't lie!" on a particular occasion, then that may very well be what would produce the best consequences in your situation. What rule and act utilitarians disagree on is what _makes right acts right_. Is it the fact that a particular act produces optimal consequences or the fact that the act conforms with a set of rules that would lead to optimal consequences if an overwhelming majority in society were to accept those rules?

THE RIGHT AND THE GOOD

Nearly all ethical theories consist of two distinct subtheories: a theory of what is good for its own sake ("the good") and a theory of what makes right acts right ("the right"). The most well-known version of utilitarianism is classic, hedonistic act utilitarianism.

This theory can be characterized as the conjunction of the claim that only happiness (well-being) is good for its own sake, where happiness is interpreted as the net balance of pleasure over pain, and the claim that an act is right if and only if no alternative act brings about more of what is good for its own sake.

Bentham and Mill discuss the concept of happiness extensively. They both propose intricate procedures for how to measure and calculate pleasure and pain. According to Bentham, an accurate calculation of how much pleasure an act brings about has to take the following four factors into account:

1. Its intensity
2. Its duration
3. Its certainty or uncertainty
4. Its propinquity or remoteness

The most valuable form of happiness is, of course, happiness that has high intensity, lasts for a long duration, is certain to occur, and is not remote. From a technical point of view, one could interpret Bentham as claiming that the *utility* of the consequences brought about by an act is a function of four variables. If so, we can represent the utility of an act's consequences by letting u(intensity, duration, certainty, propinquity) be a real-valued function of the variables listed by Bentham that assigns a number to each experience faced by a sentient being. The concept of utility is a technical concept used by utilitarians for assigning numbers to whatever it is that they claim to be valuable for its own sake, for example, the balance of pleasure over pain. Although Bentham himself did not explicitly use any numerical structure for measuring pleasure, it is plausible to think that he thought that pleasure could be represented by real numbers.

A noteworthy feature of Bentham's hedonistic theory of the good is that it is *additive*: If Anne's level of pleasure is 20 units and Bob's is 30 units, they together experience 50 units of pleasure. More generally, Bentham would say that the overall utility of a distribution of pleasure equals the sum total of pleasure produced by the act, regardless of how that sum is distributed. If you can either increase the pleasure for someone who is very well off by 100 units, or improve the situation for someone who is much worse off by 99 units, utilitarians believe that you ought to opt for the former alternative. Equality is thus of no direct moral relevance. That said, the fact that Bob is better off than Anne could of course matter if the relative differences in pleasure somehow turn out to have positive or negative side effects on someone's mental state. For instance, if Anne gets jealous of Bob because he is better off, and if the fact that Anne feels jealous makes her even worse off, then the sum total of pleasure in society would be affected, meaning that inequality is bad in this indirect sense.

Mill's theory of "the good" is different from Bentham's. Although Mill is also a hedonist, he famously distinguished between *higher* and *lower* pleasures. The kind of pleasure you enjoy when you read a complex philosophical text, or listen to good music, is worth more than the kind of pleasure you experience when you watch football, or chat with your friends on social media. Mill even claims that

It is better to be a human being dissatisfied than a pig satisfied; better to be Socrates dissatisfied than a fool satisfied. And if the fool, or the pig, is of a different opinion, it is only because they only know their own side of the question.[3]

CASE 5-3

John Stuart Mill

Figure 5.1
John Stuart Mill (1806–1873), photographed by John Watkins c. 1870.
Source: Hulton Archive, London Stereoscopic Company.

John Stuart Mill was born in 1806 in London (Figure 5.1). He was the son of James Mill, a Scottish philosopher and historian. James Mill and Jeremy Bentham, whose work on utilitarianism is discussed in this chapter, knew each other well. John Stuart was given a very comprehensive education from a very early age. According to some accounts Bentham participated actively in John Stuart's education. Mill learned Greek at the age of four and Latin at the age of eight. Within a couple of years, he had read all the ancient masters and knew almost all there was to know about math, physics, and astronomy. At the age of fourteen, Mill went to France to study abroad for a year. Because he did not accept the religious doctrines of the Church of England, he was not allowed to study at Oxford or Cambridge (which was probably a bigger loss for Oxbridge than for Mill).

Between 1823 and 1858, Mill worked as an administrator in the British East India Company. He later became Lord-Rector (president) of the University of St Andrews and member of parliament (MP). In 1866, Mill became the first MP in British history to propose that women should be given the right to vote. Some of Mill's most important works are *Principles of Political Economy* (1848), *On Liberty* (1859), *Utilitarianism* (1863), *The Subjection of Women* (1869), and *Three Essays of Religion* (1874). Mill was elected a Foreign Honorary Member of the American Academy of Arts and Sciences in 1856.

Mill is an important thinker whose writings on controversial moral and political issues continues to be read and cited today. Although many of the ideas proposed by Mill were quite radical, he was not always the first thinker to propose the ideas for which he is known. The value of his work lies in his tendency to develop important ideas with great precision and care and his admirable ambition to write as clearly as possible on controversial and sometimes provocative topics.

Discussion question: How do you think John Stuart Mill would have analyzed modern copyright laws for computer software? Do we maximize utility by enforcing copyright laws?

According to Mill, the four variables identified by Bentham do not suffice for correctly characterizing the utility of a pleasurable experience. Even if the intensity of the pleasure experienced by the pig is high; and its duration is long, certain, and not remote, it is still better to be Socrates dissatisfied. This is, according to Mill, because the quality of our mental experiences also matters. Higher pleasures matter more than lower ones. Expressed in a mathematical vocabulary, the utility function proposed by Mill is a function of five variables: intensity, duration, certainty, propinquity, and quality.

HOW SHOULD THE GOOD BE DISTRIBUTED?

Utilitarianism is a special version of a broader class of *consequentialist* ethical theories. All consequentialists believe that the moral rightness and wrongness of our acts depend on nothing but consequences. However, advocates of different versions of consequentialism disagree on exactly how the moral properties of our acts correlate with their consequences. Utilitarians claim that an act is right just in case the sum total of utility (happiness, well-being) produced by the act is maximized, but this aggregation principle is by no means accepted by all consequentialists.

Henry Sidgwick was the first author to discuss the distinction between utilitarianism and other versions of consequentialism in detail. He pointed out in *The Methods of Ethics* that ethical egoism is also a consequentialist theory. Ethical egoists believe that an act is right if and only if the consequences for the agent performing the act are optimal, no matter what the consequences are for others. Just as in the utilitarian theory, the rightness and wrongness of an act thus depends exclusively on consequences. Intentions, duties, or rights do not matter. Ethical egoism should not be confused with *psychological egoism*, that is, the claim that people do in fact do what they believe to be best for them. This is an empirical hypothesis about human behavior. It is not an ethical theory.

Although many people find ethical egoism implausible, Sidgwick and other prominent consequentialists take egoism to be a serious alternative worth considerable attention. Why should we care about others? Do we really have any ultimate or nonderivative reason to care about anyone else than ourselves? This question is no doubt very difficult to answer.

Other consequentialist theories make different claims about the correlation between consequences and moral properties. For instance, consequentialist *egalitarians* believe that an act is right if and only if the well-being produced by the act is distributed equally in society. The key claim that all consequentialists endorse is that whether something has some moral property depends only on consequences.

Traditionally, consequentialists have been concerned with the moral evaluation of *acts* (or sets of acts). However, it has been proposed that other entities—such as political institutions, laws, desires, dispositions, emotions, and even beliefs—have moral properties that depend solely on consequences. The most extreme view of this type is *global consequentialism*. Global consequentialists believe that we are morally obliged to see to it that we all have the emotions, desires, and dispositions that lead to the best consequences. This view of course presupposes that we can somehow decide, at least indirectly, which emotions, desires, and dispositions to have.

Non-consequentialist ethical theories assign (at least some) weight to entities other than consequences. Hybrid theories, which assign moral weight to consequences as well as to something else (such as intentions), are thus classified as non-consequentialist theories.

Consequentialist theories can also be divided into one-dimensional and multidimensional views. One-dimensional consequentialists believe that an act's rightness depends on a single aspect or variable, such as the sum total of well-being it produces, or the sum total of priority- or equality-adjusted well-being. Multidimensional consequentialists believe that an act's rightness depends on more than one aspect or variable, such as the sum total of well-being *and* the degree of equality produced by the act, to directly influence the act's deontic status.

Utilitarianism, prioritarianism, and egalitarianism are well-known examples of one-dimensional theories. All these theories take moral rightness to depend on a single moral variable. Utilitarians believe that an act is morally right if and only if the sum total of well-being it produces is at least as high as that produced by every alternative act. Prioritarians believe that benefits to those who are worse off should count for more than benefits to those who are better off, meaning that well-being has a decreasing marginal value, just like most people have a decreasing marginal utility for money. Egalitarian theories come in many different versions, but a unifying idea is that moral rightness depends on relative differences in well-being.

Multidimensional theorists believe that we have to make a genuine moral compromise between two conflicting variables. For example, if we cannot maximize both equality and the sum total of well-being, it can still be the case that both variables determine the act's moral rightness. In such a case, the alternative that scores best with respect to overall well-being is right with respect to one variable (well-being) while the other is right with respect to the other variable (equality). All things considered, no alternative is entirely right or wrong because none of the variables trumps the other. According to multidimensional consequentialists, both alternatives are, literally speaking, a little bit right and a little bit wrong.

SOME OBJECTIONS TO UTILITARIANISM

Many objections to utilitarianism, which have been raised many times over the years by its critics, seek to show that although an act would as a matter of fact bring about the greatest amount of utility, it would be morally wrong to perform that act. Here is an example: From a utilitarian point of view it would, arguably, be morally right to make cyclists legally required to wear helmets because that would reduce the number of fatalities in traffic accidents (see Figure 5.2). However, critics argue that such a paternalistic law would violate the right of the individual to freely decide what risks to assume (given that no one else is affected and the healthcare costs are covered by one's private insurance). If you value the freedom of the individual regardless of how this freedom affects the number of fatalities in traffic accidents and our overall well-being, then you have to conclude that a paternalistic law requiring cyclists to wear helmets would be morally wrong, which indicates that there is something wrong with the utilitarian theory.

For a somewhat less realistic example, we can imagine that five patients in a nearby hospital will die unless they receive new organs. A heathy patient is sitting in the doctor's waiting room. His organs happen to be a perfect match for the five patients. The doctor, who is a devoted utilitarian, realizes that he could kill the healthy patient and give the organs to the five needy patients. Because the consequences of killing one healthy person to save the five are better than the consequences of not doing so, mainstream accounts of utilitarian entail that it would be right to kill the healthy patient. This seems deeply counterintuitive.

The utilitarian response to this objection is to point out that we also have to consider the *negative side effects* of intentionally killing a healthy patient in a hospital. If we were to do so, people would lose their confidence in the healthcare system. Many people would become afraid of visiting hospitals, which would in the long run lead to

Figure 5.2
Up to 150 cyclists per year are killed in traffic accidents in the Netherlands. Should cyclists become legally required to wear helmets? From a utilitarian's perspective, this depends on the consequences for everyone affected by the law. Paternalistic laws are morally acceptable if their total consequences are optimal.   Source: iStock by Getty Images.

worse consequences. However, at the end of the day, utilitarians must, as pointed out by Philippa Foot and others, concede that it would be right to kill an innocent person if the consequences of doing so would be optimal. Perhaps the organs of the healthy patient are secretly harvested by a small team of specialized doctors (working in a secret facility) who never tell anyone about their work. If so, it seems that no negative side effects would occur, meaning that the consequences of killing the healthy patient would be optimal.

Another type of objection, which is based on similar concerns, is that utilitarianism and other consequentialist theories cannot honor the special moral boundary between persons. According to philosopher John Rawls, the consequentialist's conception of a person is "that of a container-person: persons are thought of as places where intrinsically valuable experiences occur, [with] these experiences being counted as complete in themselves. Persons are, so to speak, holders for such experiences."[4] Rawls's point is that, for consequentialists, persons are like mere containers for well-being, but the container itself has no direct moral importance. The theory allows us to freely "transfer" well-being from one person to another as long as the sum is constant. If, say, Anne were to redistribute well-being from Bob to herself by stealing his car, and Bob's loss is precisely counterbalanced by Anne's gain (and no one else is affected), then it would be morally permissible for Anne to steal Bob's car. This is counterintuitive.

A third type of objection is that the utilitarian theory is too demanding. If the only morally right acts are those that bring about the *best* consequences, then nearly all acts we perform are wrong. For instance, instead of reading this chapter, you could have been campaigning for Oxfam or worked as a banker for a while and then donated your huge bonus to the Red Cross. Those options would have had better consequences, meaning that it was wrong to refrain from doing so. Obviously, you preferred to read this text, which you knew would not bring about the maximum amount of good consequences in the world. Utilitarians must thus conclude that your act was morally wrong, which does not square well with our considered moral intuitions. This indicates that there is either something wrong with the theory or with our intuitions.

Despite all these objections, classic utilitarianism and consequentialism continue to be the most extensively discussed and researched ethical theories. A possible reason for this is that it seems obvious that consequences matter at least to *some* extent for what we ought to do. To claim that the consequences of our acts are totally irrelevant seems implausible.

ETHICAL EGOISM

Ethical egoism is a radical alternative to utilitarianism. Just like other consequentialists, ethical egoists believe that the moral rightness of an act depends exclusively on consequences. However, unlike utilitarians and other consequentialists, ethical egoists believe that it is morally right to do whatever produces the best consequences for the agent herself or himself. To put it briefly, you should do whatever makes thing go best for you, and I should do whatever makes things go best for me.

Ethical egoists stress that people we interact with are often more likely to be nice to us if we are nice to them. Therefore, the practical implications of ethical egoism are not as radical as one might think. In many everyday situations, the best outcome for oneself is achieved by behaving in the same way as non-egoists. If your acquaintances find out that you are an egoist, they are likely to punish you; so if you are an egoist, it might be in your best interest to not tell anyone about this. However, the consequences of some of the actions open to us

---

CASE 5-4

Yuzuki's Career Choice

Yuzuki is a graduating senior in electrical engineering. Her dream job is to work for Google in Silicon Valley. Four months ago she submitted her resume, but she has not yet been invited for an interview. Yuzuki is now starting to feel desperate and uncertain about her future. She asks the University Career Center to help her find another job. The friendly staff turn out to be good at their trade. A few weeks later she gets an offer from the IRS (Internal Revenue Service) to work on their thirty-year-old computer system. This is not Yuzuki's dream job, and the salary is not as competitive as in the private sector. However, since Yuzuki has no other offer, she accepts the IRS

position by calling their recruiter and confirming via email that she will start on September 1.

Five weeks after accepting the offer from the IRS, but before Yuzuki has signed the written contract, she gets a phone call from Google. They apologize for taking so long to process her application, but the good news is that they offer Yuzuki her dream job. The salary is $15,000 higher than what the IRS offered. Would it be morally permissible for Yuzuki to change her mind and accept the job at Google?

Although Yuzuki initially believes that it would be best for her to accept the offer from Google,

(Continued)

this belief may later turn out to be false. The Google job may not be as challenging and rewarding as she thought, and perhaps the high cost of living in Silicon Valley will make her large paycheck feel smaller than the somewhat lower salary offered by the IRS for working in another part of the country. Yuzuki also has to consider the fact that turning down the IRS job, which she accepted via email, may harm her reputation if word gets around. From a strict egoistic point of view, it thus seems that the facts are somewhat unclear. Although Yuzuki feels tempted to accept the offer from Google, it may turn out that what would be best for her is to honor her commitment to the IRS.

Needless to say, ethical egoism is a controversial theory. One may intuitively feel that the ethical issue at stake here is whether Yuzuki should be *exclusively* concerned with how things go for herself. When Yuzuki accepted the IRS offer, their recruiter most likely notified the other candidates that their applications had been rejected. At that point, the other applicants may have decided to accept alternative job offers, meaning that they are no longer willing or able to replace Yuzuki at the IRS. If Yuzuki decides to accept the offer from Google, then that is likely to have negative consequences *for others*.

Discussion question: Is it morally permissible to be an ethical egoist about some issues (such as job offers) but not about others? If so, how would you respond to the objection that ethical theories are by definition universal and must therefore be applicable to all choice situations?

---

are sometimes unknown or difficult to foresee, meaning that we may not be in a position to conclude with certainty what an ethical egoist should do in a particular case.

A possible reason for accepting some form of ethical egoism could be that we all seem to be socially or genetically *disposed* to be egoists. To be selfish is to be human. Therefore, there is nothing wrong with doing what is best for oneself.

Is this a convincing argument? Let us suppose, for the sake of the argument, that the premise is true. Most of us are, as a matter of fact, socially or genetically disposed to be selfish. Even if we accept this premise, it does not follow that we *ought* to be influenced by our selfish dispositions. This would be a violation of Hume's Law. To see this, imagine that you are a white man serving on the hiring committee in your company. Nine of the ten applicants are white males, just like you. The tenth applicant is a woman of color. Psychological research shows that white men are disposed to favor other white men in hiring decisions at the expense of other candidates, but this does not justify their behavior. We all agree that you and the other members of the hiring committee ought to treat all candidates fairly, no matter how you are socially or genetically disposed to behave. The same applies, it seems, to the disposition to be selfish. If you believe that you are disposed to be selfish, then this is a type of bias you ought to be aware of but not base any decisions on.

A common objection to ethical egoism is that the theory is *self-defeating*. The essence of this objection is that if all of us do what is best for ourselves, then everyone will end up in a situation that is worse from an egoistic point of view, compared to if we had not been selfish. A classic example is the prisoner's dilemma. Imagine that Alice and Bob have been arrested by the sheriff for some serious crime. The district attorney gives them one hour to decide to either confess or deny the charges. If both Alice and Bob deny the charges, and thus cooperate with each other, each will be sentenced to three years in prison. However, if one of them tries to cooperate with the other person (denies the charges) and the other does not, then the person who confesses (does not cooperate with the other prisoner) will be rewarded and get away with serving just one year. The other person will be punished with twenty-five years. If both Alice and Bob decide to not cooperate (confess), each will be sentenced to eight years. Alice and Bob are not allowed to communicate with each other. Table 5.2 gives an overview of the four possible outcomes.

Table 5.2 In the Prisoner's Dilemma, the first number in each cell represents the outcome for Alice, and the second number the outcome for Bob

		Bob	
		Cooperate	Do not
Alice	Cooperate	−3, −3	−25, −1
	Do not	−1, −25	−8, −8

The numbers −1, −25 mean one year in prison for Alice and twenty-five years for Bob, and so on. Imagine that Alice and Bob are fully rational, ethical egoists who want to minimize their time in prison regardless of what happens to the other person. Shall they confess or deny the charges?

The problem is that as ethical egoists, both Alice and Bob are rationally required to confess to the charges *no matter what the other person does*. If Bob opts for the cooperative strategy (cooperates with Alice by denying the charges), then one year in prison for Alice will be better for her than three; and if Bob decides not to cooperate, then eight years in prison is better for Alice than twenty-five. Furthermore, because the situation is symmetric, Bob will reason in the same way and refrain from cooperating no matter what Alice decides to do. Therefore, both Alice and Bob will serve eight years each in prison. This is a remarkable conclusion because if Alice and Bob had been willing to take the consequences for the other person into account, each of them would have chosen to cooperate and served three years in prison, which would have been better for both of them.

Situations that are similar in structure to the prisoner's dilemma arise in many parts of society. Consider, for instance, the income tax you pay to the IRS. From each egoistic taxpayer's point of view, it might be optimal to cheat as much as possible (if this maximizes the agent's well-being and the risk of getting caught is sufficiently small). However, if all taxpayers were to cheat and refrain from paying taxes, as ethical egoism would sometimes prescribe they should, then the government would no longer be able to provide essential services to the taxpayers. The collapse of vital societal structures would arguably be worse for everyone.

Another example could be global warming. For each small- or medium-sized nation, it might be best to refrain from reducing greenhouse gases no matter what other nations do. The effects on the climate of emissions from any single small- or medium-sized nation is negligible. However, when each small- or medium-sized nation does what is optimal from their point of view, the outcome will be worse for everyone. Because the effects of global warming are very negative for all of us, it would have been better for everyone to cut down on greenhouse emissions.

What has all this got to do with ethics? The key point is that there are many situations in which ethical egoists will do what is best for each individual but is suboptimal for everyone. If I do what is best for me and you do what is best for you, both of us will suffer from this and end up in a situation that is worse for both of us. It would have been better for both us to not be egoists. This shows that ethical egoism is self-defeating in the sense that if we try to achieve what this theory says is desirable, we will end up in a state the theory considers to be undesirable. That said, many people do of course not always behave egoistically. So if you are an ethical egoist, but know that sufficiently many others are not, the theory would not be self-defeating.

REVIEW QUESTIONS

1. Give a clear and precise formulation of what it is utilitarians believe make right acts right.
2. Explain the distinction between act and rule utilitarianism. Which version of the theory is most plausible?
3. Explain the distinction between actual and expected consequences.
4. What is classic hedonistic act utilitarianism, and what are the most important strengths and weaknesses of this theory?
5. What is global consequentialism, and what are the most important strengths and weaknesses of this theory?
6. John Stuart Mill proposed the following argument for utilitarianism: "No reason can be given why the general happiness is desirable, except that each person, so far as he believes it to be attainable, desires his own happiness." Is the fact that people do in fact desire happiness a reason for thinking that happiness is desirable? If so, is this argument compatible with Hume's Law? (see chapter 4).
7. Philippa Foot asks us to imagine that five patients will die unless they receive new organs from a healthy stranger. From a utilitarian point of view, it thus seems right to kill one to save five, which seems absurd. How could utilitarians respond to this objection?
8. Explain the worry that the utilitarian theory seems to be too demanding.
9. Explain the distinction between psychological egoism and ethical egoism.
10. Explain what, if anything, the prisoner's dilemma can teach us about ethical egoism.

REFERENCES AND FURTHER READINGS

Bales, R. E. 1971. "Act-utilitarianism: Account of Right-making Characteristics or Decision-making Procedures?" *American Philosophical Quarterly* 8, no 3: 257–265.

Bentham, J. 1996. *The Collected Works of Jeremy Bentham: An Introduction to the Principles of Morals and Legislation.* Oxford, England: Clarendon Press. First published 1789.

Mill, John Stuart. 1863. *On Liberty.* Boston: Ticknor and Fields. First published 1859.

Mill, John Stuart. 1863. *Utilitarianism.* London: Parker, Son and Bourn. First published 1861.

Peterson, M. 2013. *The Dimensions of Consequentialism.* Cambridge, UK: Cambridge University Press.

Rawls, John. 1971. *A Theory of Justice.* New York: Oxford University Press.

Sen, A., and B. Williams, eds. 1982. *Utilitarianism and Beyond.* Cambridge, UK: Cambridge University Press.

Sidgwick, Henry. 1884. *The Methods of Ethics.* London: MacMillan and Co.

NOTES

1. J. Bentham, *The Collected Works of Jeremy Bentham: An Introduction to the Principles of Morals and Legislation* (1789; repr., Oxford, England: Clarendon Press, 1996).
2. Ibid. 12.
3. John Stuart Mill, *On Liberty* (1859; repr., Boston: Ticknor and Fields, 1863), 14.
4. John Rawls, *A Theory of Justice* (New York: Oxford University Press, 1971), 17.

CHAPTER 6

Duties, Virtues, and Rights

In this chapter, we discuss ethical theories that *do not* take an act's moral properties to depend solely on consequences. Just like utilitarianism and ethical egoism, non-consequentialist theories seek to answer a fairly general question: What makes right acts right and wrong ones wrong? Many ethical issues faced by engineers can, arguably, be solved without knowing the answer to this. However, as explained in chapter 5, knowledge of ethical theories often help us improve our understanding of real-world issues faced by engineers. By learning *why* something is right or wrong, we become better equipped to deal with ethical problems, even if warranted moral conclusions can sometimes be reached without invoking any ethical theory.

One of the most influential non-consequentialist theories is duty ethics, introduced by Immanuel Kant (1724–1804) in *Groundwork of the Metaphysics of Morals*. According to Kant, an act's rightness depends on the *intention* with which it is performed. Acts that are performed with a good intention are right, whereas acts performed with bad intentions are wrong.

The second theory is virtue ethics. Virtue ethicists believe than an act is right if and only if it would have been performed by a fully virtuous agent. This is an ancient theory advocated by, among others, Aristotle (384–322 BC). The virtue ethicist's criterion of moral rightness cannot be applied to real-world cases before we know what characteristics "fully virtuous" agents have. Aristotle listed four virtues he believed every moral agent needs to acquire: prudence, temperance, courage, and justice. These so called *cardinal virtues* can be supplemented with professional virtues, some of which are of particular importance to engineers.

The third and final non-consequentialist theory takes the notion of *rights* as its point of departure. Legal and political thinkers have proposed many alternative notions of rights, but two particularly influential accounts are John Locke's (1632–1704) *Two*

Treatises of Government and Robert Nozick's (1938–2002) *Anarchy, State, and Utopia.* Nozick argues that rights can be thought of as moral "side constraints." For instance, if you own a piece of land, then I am not allowed to build a house on your land without your permission. Your ownership right to your land creates side constraints for what other people are permitted to do.

KANT'S DUTY ETHICS

According to Kant, only acts that are performed *from a sense of duty* are morally right. If you merely act *in accordance with duty*, that is, you perform the right act with the wrong intention, then your act has no genuine moral value, even if you end up *behaving* in the same way as someone acting from a sense of duty. Consider, for instance, an engineer who refrains from overcharging a somewhat naïve client who he knows could easily be fooled. If the engineer decides to not overcharge the naïve client because he believes that could harm his reputation and have bad consequences for his firm in the future, this would, according to Kant, be an example of an agent who acts *in accordance with duty* but not *from a sense of duty.* The only morally right option for the engineer is to not overcharge the naïve client *because* he wants to act from a sense of duty.

In chapter 1, we discussed the Citicorp case and the press release issued by William LeMessurier and his associates about the need to reinforce the wind-bracing system of the Citicorp tower in New York. The press release explicitly stated that there was "no danger" to the public, which was an outright lie. As you may recall, the risk that the tower would topple in any given year was roughly equal to the probability that the king of spades would be drawn from a regular deck of cards. In September 1978, Hurricane Ella was just hours away from hitting the tower. Did LeMessurier lie to the public out of a sense of duty, or did he merely act in accordance with duty, or perhaps contrary to duty? That is, did LeMessurier act with a good intention? Let us suppose, to simplify the example somewhat, that we know that Citicorp and LeMessurier lied *because* they wanted to protect the health, safety, and welfare of the public. In many everyday contexts, this would count as a good intention, meaning that the lie could be excused. However, Kant would disagree with this. He proposed a more demanding analysis of what counts as a good intention. Let us unpack Kant's theory step by step. His point of departure can be formulated as follows:

> An act is right if and only if it is performed with a good intention.

By "good intention" Kant means "good will." According to Kant, acts performed with "a good will" are right under all circumstances and in all situations:

> Nothing can possibly be conceived in the world, or even out of it, which can be called good, without qualification, except a good will.[1]

Kant was aware that utilitarians would disagree with him about this. Utilitarians would, of course, claim that the only thing that is "good without qualification" is happiness or well-being. Kant's response to this utilitarian objection is that "the overall wellbeing and contentment with one's condition that we call 'happiness' create pride, often leading to arrogance, if there isn't a good will to correct its influence on the mind."[2] (Utilitarians would reply that negative side effects of happiness, such as "pride, often leading to arrogance," are acceptable *as long as everyone remains happy*. From a utilitarian perspective, arrogance is not bad in itself.)

Kant distinguishes between two types of imperatives or commands. *Hypothetical* imperatives take the form of if-then clauses: *If* you wish to achieve some goal x, *then* perform action y for reaching x. Here are some examples: (1) "if you wish to eat good pasta, then your intention should be to dine in an Italian restaurant"; (2) "if you wish to build a safe railroad bridge, then your intention should be to follow the FRA Bridge Safety Standards"; (3) "if you wish to pass the exam in engineering ethics, then your intention should be to read this textbook carefully." Hypothetical imperatives differ in fundamental ways from *categorical* ones. A categorical imperative applies to *everyone* under *all* circumstances, regardless of what one's goal happens to be. According to Kant, only categorical imperatives are relevant in discussions of ethics. What intention you should have *given* that you want to reach a certain goal is ethically irrelevant.

CASE 6-1

Immanuel Kant

Immanuel Kant is one of the most influential philosophers of all time (see Figure 6.1). He was born in Königsberg in Prussia in 1724 and died in the same city in 1804. Although Kant grew up in a conservative Christian home, and even though many of his ethical views are compatible with Christian ethics, Kant believed that moral conclusions should be based on *reason* instead of religious belief. As rational human beings, we have the ability to figure out what is right and wrong by just exercising our intellectual powers, without consulting religious authorities or collecting a vast amount of information about the consequences of our actions.

Kant did not merely develop an ethical theory. His books on the foundations of mathematics and the theory of knowledge (epistemology) are widely discussed by contemporary philosophers. Kant developed almost all his ideas toward the end of his life. He published very little during the early years of his career; and at that time, he was not particularly influential. If he had lived today, he would probably not have been awarded tenure at a modern research university. However, during what we might describe as a midlife crisis, Kant "was awaken from his dogmatic slumber" and wrote a

Immanuel Kant.

Figure 6.1
Immanuel Kant (1724–1804) believed that an act is right just in case it is performed with the right type of intention.
Source: Shutterstock.

large number of books in a relatively short time. All these texts form a unified body of ideas, based

(Continued)

on an appeal to *reason* as a method for gaining philosophical knowledge.

Kant is, among other things, famous for his distinction between knowledge we gain *a priori* (by merely using reason) versus knowledge we gain *a posteriori* (by using our experience), as well as for his distinction between *analytic* and *synthetic* judgments. Kant thought that analytic judgments are true or false in virtue of their meaning ("All bodies are extended"), whereas the truth of a synthetic judgment cannot be established by analyzing its meaning ("All bodies are heavy"). Somewhat surprisingly, Kant believed that mathematical judgments, such as "seven plus five equals twelve," is a synthetic judgment we know a priori.

Discussion question: Do you think Kant would classify a moral judgment such as "I will borrow some money from a colleague by making a false promise to pay it back later, so I get some money without having to work" as an analytic or synthetic judgment? Do we know whether this judgment is correct a priori or a posteriori?

THE UNIVERSALIZATION TEST

Kant proposed three formulations of the categorical imperative, that is, commands that apply to *everyone* under *all* circumstances. He claimed that all three versions are equivalent and correctly characterize the intention (or will) of an agent acting out of a sense of duty. According to Kant, every practical conclusion that can be reached with one formulation of the categorical imperative can also, in principle, be reached with any of the two other formulations. In that sense, it does not matter which formulation one prefers. However, it may sometimes be easier to figure out what a Kantian would think about a particular case by considering one formulation rather than another, so it is worth discussing all three.

According to the first formulation, intentions are good (in the sense that the agent acts from a sense of duty) if they can be *universalized*. To test whether an intention passes the universalization test, we can ask ourselves the following question: Can a rational agent wish that everyone act with such-and-such intentions under all circumstances? If the answer is no, it is morally wrong to act with such-and-such intention. Kant formulated this criterion of moral rightness as follows:

> *The universalization test*
> Act only according to that maxim whereby you can, at the same time, will that it should become a universal law.[3]

A *maxim* is the subjective rule that governs the intention with which the agent is acting. It can be formulated by stating one's (motivating) reason for doing something. Here are some examples: (1) "I will cheat in the final exam so I get an A without having to study"; and (2) "I will borrow some money from a colleague by making a false promise to pay it back later, so I get some money without having to work." A *universal law* is no legal law: it is a cumbersome way for Kant to ask the reader to imagine a world in which everyone follows the same maxim.

The application of the universalization test to a particular case can be broken down into a sequence of separate steps.

1. Formulate a maxim *m* that summarizes your reason for performing action *a*.
2. Imagine a world in which maxim *m* is a universal law, known and followed by everyone, that governs the actions of all rational agents performing action *a*.
3. Ask yourself whether it is *conceivable* that *m* could be a universal law that governs the actions of all rational agents performing action *a*.

If it is *inconceivable* that maxim *m* could become a universal law, then the process stops at the third step. For an example of such a maxim, imagine that you borrow some money from a colleague and promise to pay it back later, without having any intention to actually do so. If *everyone* were to make false promises, and everyone knew this, then promising would become meaningless. You would not trust my promises, and I would not trust yours. A universalized intention to make false promises would thus make all promises self-defeating. We can see this as a form of practical contradiction.

It is not the bad consequences of universalizing the maxim that matter. Kant was not a rule utilitarian; it is irrelevant whether the net amount of pleasure or well-being would increase when the maxim is universalized. Making promises in a world in which everyone follows the maxim of making false promises is not *conceivable*, no matter how much or little pleasure or well-being that would generate. Here is another example of a maxim that is self-defeating in a similar sense. Imagine that you are about to download a movie from the Internet that is protected by copyright without respecting the copyright laws. If all of us were to violate the copyright laws all the time, then the concept of copyright would become meaningless. It is not *conceivable* that someone owns the copyright in a meaningful sense if the copyright is constantly violated. Therefore, it is wrong to violate copyright laws.

Maxims that do not pass the conceivability test at the third step give rise to *perfect duties*. Your duty to not make promises that you do not intend to keep, as well as your duty to not violate copyright laws, are examples of perfect duties. However, not all morally wrong actions are detected by the conceivability test. This is because in addition to the perfect duties, we also have many *imperfect* duties. Imperfect duties are identified by proceeding to the fourth and final step:

> **4.** Ask yourself whether you could *rationally will* that *m* becomes a universal law governing the actions of all rational agents performing action *a*.

Some actions that pass the third step will fall through at the fourth step. Kant mentions an example in which someone has the option to assist a person in distress. Imagine, for instance, that you walk by a small pond in which a three-year-old girl is about to drown. You are the only person around; and you know that unless you rescue the girl, she will die. A world in which all children in distress are left to die by passers-by is *conceivable*, so this maxim passes the third step. However, you cannot *rationally will* that all of us were to live in such a world. You, therefore, have an imperfect duty to rescue the drowning girl. Another example mentioned by Kant is a lazy student who neglects her natural talent for math. According to Kant, the student has an imperfect duty to cultivate her talent for math because she cannot *rationally will* that *everyone* who has such a talent ignores it.

Perfect duties must be fulfilled under all circumstances. They allow for no exceptions. Imperfect duties can sometimes be overridden by other duties, and the agent has some freedom to decide how to fulfill it. If, for instance, you can only rescue the drowning girl by putting yourself in great jeopardy (perhaps the pond is full of crocodiles), then your imperfect duty to rescue the girl might be overridden by your perfect duty to preserve your own life.

Imperfect duties are, in an indirect sense, evaluated by considering the consequences of adopting the universalized maxim. Whether you can *rationally will* that we all live in a world in which drowning children are left to die by passers-by seems to depend on the consequences of implementing that maxim. To maintain that what it is rational to will has nothing to do with the consequences of adopting a maxim seems implausible. Arguably, every reasonable concept of practical rationality must be sensitive to what is likely to happen to us. Although this does not turn Kant into a rule utilitarian, it is fair to say that imperfect duties are to some extent evaluated according to standards that are similar, albeit not identical, to those applicable to the rules favored by rule utilitarians.

CASE 6-2

Anne Frank and the Nazis

Anne Frank was born in a Jewish family in Frankfurt in 1929. When the Nazis seized power, she fled together with her family to the Netherlands. It soon became clear that they would not be safe there. In May 1940, the Nazis invaded the Netherlands; and in July 1942, Anne Frank and her family went into hiding in a concealed room in a building in central Amsterdam (see Figure 6.2). Two years later, the Frank family was betrayed and caught by the Nazis. Anne Frank died in Bergen-Belsen in February or March 1945, a few weeks before the concentration camp was liberated by the British army on the 15th of April, 1945. After the war, her father Otto read Anne's diaries and decided to publish her experiences of the war to honor her memory. Today Anne Frank's diaries serve as a timeless reminder of the cruelty of the Nazi regime.

Let us suppose, for the sake of the discussion, that the helpers of the Frank family in Amsterdam were devoted Kantians. How should they, as Kantians, have responded if a Nazi officer had knocked on the door and asked, "Is Anne Frank hiding in this building?" The universalization test implies that all of us have a perfect duty not to lie. A world in which everyone lies all the time is *inconceivable*, in the sense that no meaningful communication would be possible. The maxim "lie when the consequences of doing so are good" would contradict itself. Therefore, the person who answers the door must, according to Kant, tell the truth: "Yes, Anne Frank is here." The fact that this answer would likely have caused her death would have been morally irrelevant. The perfect duty to not lie trumps the imperfect duty to assist people in great distress.

Figure 6.2
The Anne Frank (1929–1945) house in Amsterdam. Anne Frank died in Bergen-Belsen in February or March 1945, a few weeks before the concentration camp was liberated on April 15th, 1945.
Source: iStock by Getty Images.

Kant was aware of this somewhat controversial implication of his theory. It had been pointed out to him by, among others, Maria von Herbert in a series of letters. Kant's response was that the person who answers the door does nothing wrong by being honest to the murderer. Each person is her own moral agent. According to Kant, it is the murderer (the Nazi) who wants to kill the innocent victim who is guilty of moral wrongdoing, not the person who leads the murderer to the victim.

A problem with Kant's response seems to be that we can imagine other scenarios in which the only way to save the life of an innocent person is to lie, but in which there is no other person to blame for the murder. For a somewhat unrealistic example, imagine that some very complex software system for air traffic control has been accidentally programmed such that unless you deliberately lie in a lie detector test, in which you are asked an innocent question about what you did last night, the machine will send a signal to the system that will make two large aircraft collide in midair. In this version of the example, there is no "external murderer" to blame. The only thing that determines whether innocent people will die is whether you lie about something fairly unimportant.

Discussion question: Is it always morally wrong to lie, under all circumstances, even if you tell a "white lie" that saves the lives of hundreds of people? If not, what role can, and should, consequences play in duty ethics?

MEANS TO AN END AND RESPECT FOR PERSONS

According to Kant's second formulation of the categorical imperative, it is wrong to treat others as *mere means* to an end. Here the key idea is that we should show respect for human beings as persons:

> *The end-in-itself test*
> Act in such a way that you treat humanity, whether in your own person or in the person of any other, never merely as a means to an end but always at the same time as an end-in-itself.[4]

As an illustration, consider the owner of a cotton plantation who employs slave labor. The owner is treating his slaves as mere means to an end because the relationship between the owner and the slave is purely instrumental. The owner is not respecting the slave as an end in itself. All the slave owner cares about is the cotton produced by the other person.[5]

Note that Kant thinks it is wrong to treat people as a *mere* means to an end. It is, of course, permissible to treat someone as a means to an end as long as he or she is *also* treated as an end in itself. If you take a taxi from the airport to your hotel, the taxi driver and you treat each other as a means to an end. You use the driver as a means for getting from one location to another; the driver uses you as a means for making money. There is nothing wrong with this if you and the driver *also* show respect for each other as human beings and treat each other with dignity. In your role as customer, you must behave well and not be arrogant or take the service for granted. The driver should make sure that you reach your destination safely and not overcharge you. As long as these criteria are met, you are not treating each other as mere means to an end. Compare this to the case of the slave owner. Although slaves can of course be treated in many different ways, it seems odd to think that the slave owner could treat his "property" with the degree of dignity and respect required by the categorical imperative.

Kant's third formulation of the categorical imperative is somewhat less transparent than the others. It focuses on protecting the autonomy of the individual and respecting them as rational human beings:

The respect-for-persons test
Every rational being must so act as if he were through his maxim always a legislating member in the universal kingdom of ends.[6]

What does it mean to be a "legislating member" of a "kingdom of ends"? A common interpretation is that we should imagine a hypothetical situation in which a group of fully rational and autonomous agents sit down to agree on a constitution for a newly founded nation. Think, for instance, of the Founding Fathers, who signed the US Constitution in 1787 just two years after Kant published *Groundworks of the Metaphysics of Morals* in 1785. If the Founding Fathers had been fully rational and autonomous, and insusceptible to all forms of biases and irrelevant interests, what would they then have agreed on? It seems that nearly all of the rights and freedoms listed in the Constitution would indeed have been adopted by the legislating members in a universal kingdom of ends.

Several commentators have pointed out that the categorical imperative has much in common with the golden rule, which has been adopted in one form or the other by nearly all major religions. The golden rule states that you should, "Do unto others as you would have them do unto you." This idea is vaguely similar to the first formulation of the categorical imperative. However, a crucial difference is that while the categorical imperative is a *structural* criterion on maxims, the golden rule is only applicable if one knows how one wants other people to treat oneself. How I should treat you will thus depend on my own preference. Here is a somewhat oversimplified example that illustrates the problem. I know that I prefer coffee to tea, and that you prefer tea to coffee. I thus prefer that you serve me coffee. However, according to a strict application of the golden rule, I would have to serve you coffee even though I know you prefer tea: "Do unto others as you would have them do unto you!" A possible response to this could be to restrict the scope of the golden rule to preferences that you and I share, such as the preference to get what we want, or be asked about our preference before being served. However, it is not clear if any such universally shared preferences exist.

CASE 6-3

Airbag Recalls and the Duty to Inform

According to an article in the *New York Times* published in February 2016

> Honda and Fiat Chrysler will recall about five million vehicles worldwide to fix a defect in an airbag component known for years but left unaddressed. Continental Automotive Systems, the German supplier that manufactures electronic components that control car airbags, has been aware of a defect in some units since January 2008. [. . .] Semiconductors inside the unit could corrode, causing the airbags to deploy inadvertently or fail to deploy at all, Continental said. Once Continental knew of the problem, it informed automakers, said Mary Arraf, a spokeswoman for the supplier. She said it was up to carmakers to issue a recall. Continental then quietly fixed the problem, adopting remedies by March 2008. It did not alert regulators

of the defect at the time. Automakers have linked at least nine injuries to the defect.[7]

How should a Kantian duty ethicist judge Continental's decision to inform the automakers of the defective airbags but not the regulatory authorities or the car owners? According to Kant's universalization test, Continental did not have any *perfect* duty to inform anyone. It is *conceivable* that no one ever informs anyone about safety issues. No contradiction arises if we universalize the maxim "do not inform anyone about your defective product." However, we cannot *rationally will* that everyone follows that maxim. Therefore, Continental arguably had an *imperfect* duty to inform someone about the defective airbag (see Figure 6.3).

If so, was it sufficient to just inform the automakers? Or did the directors of Continental have an imperfect duty to inform the regulatory agency, the National Highway Traffic Safety Administration (NHTSA), and the car owners? In principle, it should be possible to answer this question by applying the fourth step of Kant's universalization test; but doing so in this and other similar cases can be difficult. It is not always easy to know what we can "rationally will." Perhaps it is rational to will that one receives as much information as possible. But perhaps it is also rational to will that we as car owners are kept uniformed until it has been determined how severe a potential safety issue is and how it can be fixed. If we receive too much information about potential hazards, there is a risk that we ignore all the safety warnings we receive.

When the universalization test fails to give a clear answer to an ethical question, it can be fruitful to open the Kantian toolbox and apply the other tests. In this case, the end-in-itself test seems to

Figure 6.3
Did Continental have an imperfect duty to inform the regulatory agencies of the defective airbags?
Source: iStock by Getty Images.

be mute. Arguably, no one was treating anyone else as a mere means to an end. However, the respect-for-persons test indicates that Continental did actually have a duty to alert the regulatory authorities around the world as well as the affected car owners. Treating someone as an autonomous agent requires that he or she has access to all the information that is relevant for a decision. In this case, both the NHTSA as well as the car owners are autonomous decision makers, so Continental therefore had a duty to provide them with all the information they could reasonably anticipate would be relevant for an autonomous decision-making process.

Discussion question: According to Continental, it was up to carmakers to issue a recall, and there was no need to alert regulators about the defective airbags. Is this moral judgment compatible with the golden rule? That is, can we accept this conclusion if we believe we should treat others as they want to be treated by us?

VIRTUE ETHICS

Virtue ethics is an ancient ethical theory advocated by, among others, Aristotle (384–322 BC) in his *Nicomachean Ethics* and the Catholic priest Thomas Aquinas (1225–1274 AD). The Chinese thinker Confucius formulated somewhat similar ideas hundreds of years earlier, around 500 BC, but his version of the theory has been less influential among Western thinkers.

A virtue can be defined as a *character trait* or *act disposition*. Aristotle claims that every agent who wishes to live a good life needs to acquire four especially important virtues, which modern theorists refer to as *cardinal virtues*: prudence, temperance, courage, and justice. These character traits are virtues because agents who have them are disposed to behave in what will always turn out to be the morally correct way in moral choice situations; and the four character traits are cardinal virtues because they are relevant for everyone in all choice situations. According to Aquinas, we should also add three theological virtues to Aristotle's list: faith, hope, and love. The cardinal virtues proposed by Confucius are somewhat different, but he agrees with Aristotle and Aquinas that some virtues are more fundamental than others.

For Aristotle, Aquinas, and Confucius, ethics is not primarily about distinguishing right acts from wrong ones. The *most important* ethical question for virtue ethicists is how we ought to live, not what we ought to do in particular situations. What makes a human life go well, and what habits are important if we are to flourish as humans?

Some modern virtue ethicists think the cardinal virtues mentioned by Aristotle, Aquinas, and Confucius should be supplemented with *professional virtues*. Professional virtues vary from profession to profession. For example, unlike professional magicians and fiction writers, engineers should arguably be honest to their clients. If so, honesty would be a professional virtue for engineers but not for professional magicians and fiction writers.

What determines an act's rightness or wrongness according to virtue ethicists is the character traits or act dispositions of the agent who performs the act rather than the act's consequences or the agent's intention. Consequences and intentions are relevant only in so far as the virtuous agent has reason to care about them. Someone guided by prudence, temperance, courage, and justice would typically not perform acts with disastrous consequences or with bad intentions; but the consequences or intentions do not ultimately make right acts right and wrong ones wrong.

The best way to understand why virtues are so important according to Aristotle is to consider cases in which virtue is lacking. Think, for instance, of a cruel engineer who deliberately designs a new airbag for the automotive industry in a way that will cause unnecessary harm to users (by intentionally omitting some relatively cheap safety feature). A comprehensive moral evaluation of the evil initiated by this engineer will be incomplete until we consider the engineer's character traits. To put it briefly, this engineer was *cruel* and *insensitive* to human suffering. The fact that the engineer exhibited these character traits tells us something important about what he did and why, which cannot be fully captured by merely evaluating each and every individual act. The negative value of the whole is greater than the sum of its parts.

It is, fortunately, rare to come across professional engineers whose character traits are as morally problematic as those just mentioned. Many engineers are praiseworthy persons with character traits that tend to lead to responsible and well-considered moral decisions. That being said, Aristotle's theoretical point is nevertheless important. If we were to *merely* evaluate individual actions taken by engineers or others, we run the risk of overlooking morally relevant features of a case. Consider, for instance, the way William LeMessurier handled the crisis with the Citicorp building discussed in chapter 1. When he found out that the Citicorp tower might topple in a moderate hurricane, he took a sequence of actions, nearly all of which seem to have been right: he informed the owners of the building about the problem, he worked out a plan for repairing and strengthening the tower, he appointed a team of weather forecasters to monitor the wind 24/7, and he approached the City of New York to work out a contingency plan in case an emergency evacuation would be needed. The only action LeMessurier could be criticized for is, it seems, the decision to not provide accurate and objective information to the public.

Virtue ethicists maintain that the best way of evaluating the Citicorp crisis is to identify the character traits that governed LeMessurier's actions. It seems that LeMessurier was concerned about the health, safety, and well-being of the public. He selflessly jeopardized some of his professional prestige to ensure that no one was hurt or killed. He agonized over the case, and he did not try to conceal the problem to his client; he was on the contrary very honest in his communication with Citicorp. When he realized his mistake, he admitted it without further ado. All this supports the conclusion that, overall, LeMessurier was a virtuous engineer who managed the Citicorp crisis well.

In Aristotle's theory, virtues are defined as character traits that are firmly established or entrenched in the agent's personality. If you, for instance, have a habit of donating 10 percent of your income to charity, and do so every month for several years, then that indicates that you are a generous person. Your disposition to make donations to charity tells us something important about who you are as a person. Compare this to someone who starts his day with a cup of coffee every morning and consistently chooses cappuccino instead of filtered coffee. The habit of choosing cappuccino is well entrenched in, but it is not a character trait that tells us anything important about your personality: meaning that this act disposition does not count as a virtue. If you were to change your

---

CASE 6-4

Dr. Elisabeth Hausler, a Virtuous Engineer

Dr. Elisabeth Hausler, a first-generation college student, graduated from the University of Illinois–Urbana Champaign in 1991 with a BS in Engineering. After working for a couple of years in the engineering consulting industry, she returned to academia for graduate studies. In 2002, she obtained a PhD in Civil Engineering from Berkley with a dissertation on geotechnical engineering.

As a Fulbright Scholar at the Indian Institute of Technology in Bombay, Dr. Hausler witnessed the effects of the 2001 earthquake in Gujarat, India. The earthquake killed between 15,000 and 20,000 people and destroyed over 400,000 homes. As an expert on geotechnical engineering, Dr. Hausler realized that relatively simple modifications to existing buildings in developing nations could dramatically reduce damage caused by earthquakes, hurricanes, and other natural disasters. She knew she had the right qualifications to do something to help solve this global problem.

In 2004, Dr. Hausler founded Build Change, a non-profit organization that designs disaster resistant houses and schools in developing countries and trains local engineers and homeowners to redesign

their buildings. Between 2004 and 2016, Build Change was instrumental in retrofitting 48,000 homes and schools. This increased the safety and future well-being of more than 250,000 people in Indonesia, China, Haiti, Colombia, Guatemala, and the Philippines. As CEO of Build Change, Dr. Hausler oversees a staff of about 150 people, most of whom are based in the developing countries in which the organization is active.

Dr. Hausler's achievements and choice of career path suggests she might be a highly virtuous engineer: she has *selflessly* and *generously* chosen a professional path that enables not only herself to *flourish* as a geotechnical engineer, but also enables thousands of others to reach their goals by living and studying in a safe environment. In her mission statement for Build Change, Dr. Hausler lists several virtues and values promoted by the organization: *technical competence, equality, curiosity, urgency, tenacity, optimism.* Dr. Hausler has been awarded numerous prizes and awards for her work with Build Change.

Discussion question: Are all or some virtues displayed by Dr. Hausler professional virtues all engineers ought to acquire? Can you think of some additional virtues that might be relevant for engineers?

preference tomorrow, that would not affect your personality. Whether something is a candidate for being a virtue thus depends on what that habit or act disposition reveals about one's personality and how it shapes one's character in ways that matter.

Not every firmly established character trait is a desirable one. The character traits of Hitler and Stalin were firmly established but highly undesirable. According to Aristotle, we can determine whether a character trait is desirable or not by considering whether it contributes to *eudaimonia*. The term eudaimonia is widely used in ancient Greek philosophy and is often translated as happiness or human flourishing. It is worth stressing, however, that the type of happiness captured by the term eudaimonia differs in significant respects from the utilitarian understanding of happiness. To flourish as a human being, you need moral as well as intellectual skills; but if you manage to reach eudemonia, you may not always be happy in the utilitarian sense. Eudaimonia is a broader concept of happiness than the narrow notion discussed by utilitarians.

Aristotle distinguishes between two broad categories of character traits, which he believes are necessary for reaching eudaimonia: moral virtues and intellectual virtues. As mentioned at the beginning of the chapter, the four most important moral virtues, called cardinal virtues, are

- Prudence
- Temperance
- Courage
- Justice

The extent to which an agent embodies a virtue varies in degree, but Aristotle does not think we should maximize virtue. We should rather strive for the desirable middle between deficiency and excess. Aristotle calls this the *doctrine of the mean*. Consider courage, for instance. Some persons have too much courage. If you are the only soldier left on the battlefield and continue the war until you get shot by the enemy, you have too much courage. It would have been better to surrender or retreat. Another example is the virtue of justice. If your employer makes you pay one cent too much for your parking spot at work, and you sue your employer because of this, you are too concerned about justice. Just as people who totally lack courage and justice have too little of those character traits, the soldier on the battlefield and the employee in the engineering firm are overly concerned with courage and justice.

In addition to the moral virtues identified by Aristotle, he also proposes a number of *intellectual* virtues he claims to be essential for a good life, such as wisdom, reason, and practical knowledge. Aristotle's point is that it is insufficient to get one's moral virtues right. Without the appropriate intellectual virtues, we are often unable to identify the morally right action. LeMessurier's plan for reinforcing the steel structure in the Citicorp tower is a good example. Without intellectual wisdom, reason, and practical knowledge, it would have been impossible for LeMessurier to come up with a technically adequate solution to the problem. So without these *intellectual* virtues, he would have been unable fulfill his *moral* obligation to strengthen the tower.

SOME OBJECTIONS TO VIRTUE ETHICS
Virtue ethics is sometimes criticized for giving moral decision makers too little action guidance. To say that an act is right if it would have been performed by a fully virtuous agent is not very helpful for those of us who are not (yet) fully virtuous. How could, for

CASE 6-5

Aristotle

Aristotle (384–322 BC) was born in the northeastern part of Greece (Figure 6.4) in a village outside modern-day Thessaloniki. At the age of eighteen, he moved to Athens to study with Plato at the world's first university, the Academy. He remained there for nineteen years. During that time, he wrote books that lay the foundations for physics, logic, biology, ethics, and metaphysics. Many of Aristotle's scientific claims turned out to be false, but he is still remembered as the first scientist. Unlike anyone before him, he made *systematic* investigations of a wide array of topics and gave *explicit arguments* for his claims. The fact that he was wrong about many things is arguably less important because he set the agenda for future research.

When Plato died, and Aristotle was not appointed director of the Academy, in a classic example of nepotism (the job went to Plato's nephew Speusippus), Aristotle traveled for a while in the Greek islands. He married Pythias, with whom he had a daughter. Four years after he left Athens, he was offered a post as private tutor of Alexander the Great in Macedonia.

After some years as private tutor, Aristotle returned to Athens in 335 BC to found his own university, the Lyceum. He taught there for twelve years. During this period, his wife Pythias died and he remarried Herphyllis, with whom he had a son and a daughter. He named his son Nichomachus after his own father. (Many historians believe that Aristotle's book on virtue ethics was dedicated to his son rather than his father, although this is somewhat uncertain.) It was during his second stint in Athens that he wrote many of his most influential books. Unfortunately, many of those texts, including all of his dialogues, have been lost. The texts that remain are primarily his lecture notes for

Figure 6.4
Many of Aristotle's best texts, including his dialogues, have been lost. The books that have been preserved are his lecture notes for the classes taught at the Lyceum in Athens. Source: Public domain.

classes taught at the Lyceum. Those texts were not intended for publication. Given how influential his manuscripts have been over the past two millennia, it seems reasonable to believe that Aristotle's texts intended for publication must have been truly amazing.

Discussion question: List some other historical figures who have influenced our thinking to roughly the same extent as Aristotle and discuss how their influence differs from Aristotle's.

instance, LeMessurier have applied Aristotle's theory if he had not been as virtuous as he actually was?

As noted, virtue ethics is first and foremost a theory about the good human life. The theory is simply not designed to answer questions about action guidance. That being said, Aristotle and his followers are keen to stress that moral virtues can be learned through experience. Imagine, for instance, that you have too little courage. For some reason, never mind why, it is morally important that you dare to give truthful answers when asked by the waiter if you like the wine in your favorite restaurant (which you do not do, because it tastes of vinegar.) By deliberately placing yourself in situations in which you need to be courageous and give truthful answers even when that is socially awkward, you gradually improve your virtue. Perhaps you decide to visit the restaurant every day for a month. The first couple of days, you may not have sufficient courage for giving truthful answers to awkward questions; but after a while, you discover that it is not actually as scary to tell the truth as you thought. By taking many small steps and giving yourself many opportunities to display courage, you eventually reach a higher level of virtue. The upshot is that even if you are not yet fully virtuous, virtue is something you can learn; and once you have improved your moral skills, you will be able to distinguish right acts from wrong ones.

In many real-world cases, it is not necessary to be fully virtuous to figure out what to do. Rather than asking "What would a fully virtuous agent have done?," it is often sufficient to ask "What would a courageous agent have done?" or "What would a prudent and courageous agent have done?" Although these questions may provide no more than a partial answer to the original question, they are often easier to answer. It is not difficult to see, not even for someone for whom virtue is lacking, that at the point in time when LeMessurier had to decide whether to inform the owner of the Citicorp tower about the problem with quartering wind loads, then the only courageous and prudential option was to take full responsibility for the problem and give full information to Citicorp.

A more fundamental objection to virtue ethics is to question the very existence of virtues. Psychological research indicates that the empirical assumption that people's actions are governed by deeply entrenched character traits, which can be cultivated and trained, might very well be false. In a series of famous experiments conducted in the 1960s, Stanly Milgram showed that surprisingly many people are willing to give (what they believed to be) very painful electric shocks to innocent strangers in laboratory settings if an authoritative person tells them to do so. Interestingly, it also turned out that people's willingness to inflict pain on innocent strangers did not depend on whether they were ordinary citizens or religious leaders or other individuals one could expect to be virtuous. Milgram found that most people were willing to give painful electric shocks to innocent strangers whenever they were told to do so in an authoritative way. This suggests that virtues, defined as stable character traits, may play no or little role for how people actually behave.

In recent years, Milgram's experiments have been criticized for being unethical. The participants did not give their informed consent to participating in a study in which they would be asked to inflict pain on others. Despite this, Milgram's work is still widely discussed by virtue ethicists. How is it possible that people we would believe to be virtuous were willing to inflict pain on others just because they were told to do so by an authoritative person they had never met before?

RIGHTS

What makes rights-based ethical theories unique is that they take rights to be the *starting point* for moral reasoning. Nearly all ethical theorists believe rights have *some* role to play in moral reasoning, in one way or another; but rights-based theories put them in the center. Two of the most important books on rights are John Locke's (1632–1704) *Two Treatises of Government* and Robert Nozick's (1938–2002) *Anarchy, State, and Utopia*. Nozick argues that rights are best conceived as moral *side constraints*. If you, for instance, own a piece of land, other people are not allowed to build a house on your land without asking for permission. Your ownership right to your land creates side constraints on what other people are permitted to do to you. In principle, they can do whatever they want as long as they do not violate your rights or those of others.

The least controversial definition of a right is to say that a right to something is a *legitimate claim* to that something. Consider, for instance, the right to free speech. According to this definition, the right to free speech is a legitimate claim to free speech. Not all legitimate claims must always be honored. In some cases, we may, for instance, have to balance conflicting legitimate claims against each other to determine what to do.

Advocates of rights-based theories distinguish between positive and negative rights. If you have a negative right to something, other people have an obligation to *refrain* from doing that something to you. If you, for instance, own a car, you have a negative right to not have your property stolen. Positive rights differ from negative ones in that other people have an obligation to see to it that you *actually get* whatever is entailed by your positive rights. According to the Universal Declaration of Human Rights, "Everyone has the right to seek and to enjoy in other countries asylum from persecution."[8] This is a positive right because it stipulates that all countries have an obligation to offer asylum to refugees.

Moral rights are sometimes confused with legal rights. The mere fact that some legal system grants you a legal right to something does not by itself entail that you have a moral right to that thing. Legal rights do not generate moral rights, at least not without additional premises. Consider, for instance, the Great Internet Firewall in China (cf. chapter 15). This is an Internet censorship and surveillance project controlled by the ruling Communist Party. By using IP blocking, DNS filtering and redirection, and URL filtering, the Chinese Ministry of Public Security blocks and filters access to information deemed to be politically dissident or "inappropriate" for other reasons. As part of this policy, the Google search engine was banned in China in 2010, as was the use of Gmail and other Google products.

Similar surveillance systems are in place in other countries. For instance, in some countries in the Middle East, websites expressing politically dissident opinions are systemically blocked by the government. This includes major media websites, such as BBC.com and other public service broadcasters. According to the local laws in China and certain countries in the Middle East, the citizens of those countries do not have any *legal right* to access websites with politically dissident information. However, it seems clear that all citizens in all countries have a *moral right* to access media websites regardless of whether they have any legal right to do so.

Philosophers have proposed two fundamentally different accounts of the origin and nature of moral rights. Some scholars argue that rights are *socially constructed*, while others maintain that they exist *naturally* and independently of us. An influential example of a social constructivist view holds that rights ultimately depend on the consequences of granting people those rights. Utilitarians have, for instance, argued that we should grant people whatever rights will bring about the best consequences. So if the consequences of giving everyone free access to the Internet are better than the consequences of blocking or filtering the Internet, then we have a moral right to access the Internet freely without being stopped by the government. This moral right to access the Internet can be compared with the right to free speech. All things considered, the consequences of granting people the right to free speech are better than the consequences of not doing so. Although some people sometimes make claims that are false, misleading, or morally outrageous, the overall consequences for society of tolerating free speech are nearly always better than the consequences in which free speech is restricted. Citizens who know that what they say, think, and write is monitored and restricted by the government are less likely to be creative, innovative, and willing to contribute to the common good.

The main alternative to the social constructivist approach is natural rights theory. Natural rights theorists believe we have our rights merely in virtue of being moral agents. The existence of moral rights does not depend on the consequences of granting us these rights, or on the outcome of any social processes, or on the adoption of some legally binding document. People just have certain rights, full stop.

The most fundamental natural right identified by Locke and Nozick is the ownership right we have to ourselves. I own myself and you own yourself. Therefore, no one is allowed to murder, torture, or rape us. Moreover, according to the part of Locke's and Nozick's theory known as the *mixing theory of labor*, we own whatever we produce with our bodies, as long as that something was not previously owned by anyone else. Consider, for instance, the Vikings who settled in Iceland about one thousand years ago (see Figure 6.5). No one

Figure 6.5
Ruins of a Viking village on Iceland, settled by Vikings in 874 AD. As long as each Viking left enough land for others to claim, which they generally did, they became the owner of a previously unowned piece of land by mixing their labor with the soil. Source: iStock by Getty Images.

owned Iceland when the first Vikings arrived. One of the first settlers was Ingolfr Arnason, who settled in Iceland in 874 AD. By mixing his labor with some piece of land (not already owned by someone else, and by leaving enough land left for others), Arnason became the moral owner of the land and the carrots he grew in the field. As owner of the carrots, Arnason was free to trade them for other goods on a free market, or give away the carrots for free if he so wanted, as long as all transactions were voluntary and reached without coercion. It was up to Arnason to decide what would happen with his property, after he died. If he so wished, he could give everything he owned to his children.

The position outlined by Lock and Nozick can be summarized in the following schema:

1. Every person owns herself or himself.
2. At some point in the past, the world was not owned by anyone.
3. As long as you leave enough for others, you become the owner of a previously unowned piece of land (or other part of the world) by mixing your labor with it and thereby taking control of the area.
4. When you mix your labor with something, you become the owner of the fruits of your labor, which you are morally allowed to trade on a free market.

By applying the four principles just listed, we can, in principle, explain phenomena such as freedom of speech and unrestricted access to the Internet. You own yourself and therefore have the right to use your own person and your belongings for expressing whatever views you want (as long as you do not infringe on the rights of others). Moreover, if the government were to block or filter information you send to others, that would be a violation of your right to exchange information and goods on the free market, even if your intention is to give away your ideas or goods for free rather than sell them for money.

The same type of reasoning can be applied to intellectual property rights, including copyright and patents. For instance, as the author of this text, I have the right to compensation from you for reading it. I have mixed my labor with a set of ideas and words owned by no one and created a text that teaches you what you need to know about the natural rights tradition. If you violate this right by reading or copying my text illegally, you thereby violate my legitimate claim to compensation.

From a philosophical perspective, the most controversial premise of the natural rights theory is the mixing theory of labor. Nozick discusses the following rhetorical objection:

> Why does mixing one's labor with something make one the owner of it? Perhaps because one owns one's labor, and so one comes to own a previously unowned thing that becomes permeated with what one owns. Ownership seeps over into the rest. But why isn't mixing what I own with what I don't own a way of losing what I own rather than a way of gaining what I don't? If I own a can of tomato juice and spill it in the sea so that its molecules (made radioactive, so I can check this) mingle evenly throughout the sea, do I thereby come to own the sea, or have I foolishly dissipated my tomato juice?[9]

According to Nozick, ownership is not contagious. Just because you mix your labor with something that is owned by no one, you do not automatically become the owner of that thing.

Critics of Locke and Nozick point out that their theories are merely capable of generating negative rights. Imagine, for instance, that you are a multimillionaire. No matter how much money you have in your bank account, you will never have any obligation to share your riches with others. You *may* make donations to charity if you want to, but you have no *obligation* to do so. This seems counterintuitive.

CASE 6-6

The Trolley Problem: A Test Case For Ethical Theories

The Trolley Problem is a classic philosophical thought experiment originally formulated by Philippa Foot in 1967 and investigated further by Judith Jarvis Thomson in 1985. As such, the Trolley Problem is not directly linked to the discussion of moral rights or any other particular ethical theory. We can rather think of it as a generic example that can be applied for assessing the plausibility of any ethical theory. Here is how Thomson formulates the problem: "Suppose you are the driver of a trolley [see Figure 6.6]. The trolley rounds a bend, and there come into view ahead five track workmen, who have been repairing the track. The track goes through a bit of a valley at that point, and the sides are steep, so you must stop the trolley if you are to avoid running the five men down. You step on the brakes, but alas they don't work. Now you suddenly see a spur of track leading off to the right. You can turn the trolley onto it, and thus save the five men on the straight track ahead. Unfortunately, . . . there is one track workman on that spur of track. He can no more get off the track in time than the five can, so you will kill him if you turn the trolley onto him. Is it morally permissible for you to turn the trolley?"[10]

How do the ethical theories discussed in this and the preceding chapter handle the Trolley Problem? For the utilitarian the choice is, of course, easy. Given that all workmen are of roughly the same age, are equally healthy, and have roughly the same number of relatives who will miss them, it seems clear that the morally right choice would be to sacrifice one person to save five. Needless to say, some philosophers find this absurd and believe that there is something fundamentally wrong with the utilitarian theory because it permits this form of aggregation of moral considerations.

Figure 6.6
The brakes of the trolley car are jammed. If you do nothing, five innocent people will die. However, if you flip a switch, the five will survive but another, equally innocent bystander, will die. What should you do, and why? Source: iStock by Getty Images.

For Kantians, it is the driver's intention that matters. If you steer the trolley to the sidetrack, you use that workman as a mere means to an end, which is impermissible. To drive home this point, it is helpful to consider a slightly different version of the example, known as the "big man case": The trolley is out of control and approaches the five workmen. You are standing on a bridge above the tracks. Next to you is a big man. The only way to stop the trolley and save the five innocent workmen is to push the equally innocent big man from the bridge onto the tracks. This version of the case seems to be analogous to the first, and here it is *paradigmatically clear* that the big man is being used as a mere means to an end, which Kantians believe is impermissible.

It is far from obvious how virtue ethicists would analyze the Trolley Problem. What would a courageous, prudent, just, and temperate agent do? I leave this to the reader to think about. However, the mere fact that this question cannot be immediately answered seems to support the worry that virtue ethics is a somewhat underdeveloped and underspecified theory. In the two thousand years that have passed since the theory was first proposed by Aristotle, we have not been able to figure out exactly what follows and does not follow from the theory.

Let us finally consider what rights might be at stake in the Trolley Problem. If you believe that we have a negative right not to be killed by anyone, then this example seems to be a case in which someone's rights will be violated no matter what you do. It is not clear that it would be better to violate one person's rights instead of five, especially because that requires that you actively steer the trolley in a different direction. Five is more than one; but just as it is wrong to violate the rights of five people, it is also wrong to violate the rights of one person. Advocates of rights-based theories could, perhaps, argue that the Trolley Problem is an example of a genuine moral dilemma of the type discussed in chapter 2.

Discussion question: Can you think of some real-world problem faced by engineers that resembles the Trolley Problem? How would you handle that problem?

REVIEW QUESTIONS

1. Kant claims that "Nothing can possibly be conceived in the world, or even out of it, which can be called good, without qualification, except a good will." Explain, by using your own words, what Kant means by this. What could utilitarians say in response to Kant's claim?

2. Explain the distinction between hypothetical and categorical imperatives, and discuss why it is important for understanding Kant's theory.

3. Explain, by using your own words, how we should understand Kant's universalization test: "Act only according to that maxim whereby you can, at the same time, will that it should become a universal law." Is this a reasonable moral principle?

4. Explain the distinction between perfect and imperfect duties.

5. Is it morally permissible to lie if that is the only way to save the life of an innocent person? What would a Kantian say, and what is your own view?

6. Kant thinks it might sometimes be permissible to treat a person as a means to an end, but he insists that it is wrong to treat persons as *mere* means to an end. Construct an example (not mentioned in this chapter) that clarifies the significance of this point.

7. Discuss the similarities and differences between Kant's theory and the golden rule. Would Kant have accepted the golden rule?

8. How would Kant evaluate the actions taken by Continental when they discovered that semiconductors inside their airbags could corrode?

9. What is a virtue? Give as many examples as you can.

10. One of Anne's many habits is to drink a cup of filtered coffee every morning. Why is this habit neither a virtue nor a vice?

11. Explain Aristotle's notion of *eudaimonia*. What is the difference between *eudaimonia* and the hedonistic utilitarian notion of happiness?

12. Explain the doctrine of the mean.

13. Explain the worry that virtue ethics may not offer decision makers any action guidance. What could virtue ethicists say in response to this objection?

14. What does it mean to say that you have a moral right to something?

15. How do moral rights differ from legal ones?

16. Explain the distinction between positive and negative rights.
17. Summarize the key ideas of Robert Nozick's theory of natural rights. What is the mixing theory of labor? What does it mean to say that rights are "side constraints"?
18. Some utilitarians reject the existence of natural rights, but nevertheless believe we should respect people's rights. Explain.
19. What is the Trolley Problem and why might it be important for testing the implications of moral theories?

REFERENCES AND FURTHER READINGS

Crisp, R., ed. 2014. *Aristotle: Nicomachean Ethics*. Cambridge, UK: Cambridge University Press.

Foot, P. 1967. "The Problem of Abortion and the Doctrine of Double Effect." *Oxford Review* 5: 5–15.

Johnson, R., and A. Cureton. 2017. "Kant's Moral Philosophy." *The Stanford Encyclopedia of Philosophy*. Spring ed. Edited by Edward N. Zalta. https://plato.stanford.edu/entries/kant-moral/

Kant, I. 1956. *Foundations of the Metaphysics of Morals*. Translated by Lewis White Beck. Indianapolis, IN: Bobbs-Merrill Library of Liberal Arts. First published 1785.

Locke, J. 1988. *Two Treatises of Government*. Cambridge, UK: Cambridge University Press. First published 1689.

Nozick, R. 2005. *Anarchy, State, and Utopia*. New York: Basic Books. First published 1974.

Tabuchi, H., and C. Jensen. February 4, 2016. "Yet Another Airbag Recall Will Affect Five Million." *New York Times*.

Thomson, J. J. 1985. "The Trolley Problem." *Yale Law Journal* 94, no. 6: 1395–1415.

Timmermann, Jens. 2007. *Kant's Groundwork of the Metaphysics of Morals: A Commentary*. Cambridge, UK: Cambridge University Press.

UN General Assembly. 1948. *Universal Declaration of Human Rights*. New York: UN General Assembly.

NOTES

1. Kant, *Foundations of the Metaphysics of Morals*, trans. Lewis White Beck (1785; repr., Indianapolis, IN: Bobbs-Merrill Library of Liberal Arts, 1956), p. 396 in the Prussian Academy edition.
2. Ibid.
3. Ibid. 421 in the Prussian Academy edition.
4. Ibid. 429 in the Prussian Academy edition.
5. The view on slavery presented here is what Kantians *should* say about this topic, not the view Kant *actually defended*. Kant's own view on slavery seems to square poorly with his general theory.
6. Kant, *Foundations of the Metaphysics of Morals*, p. 438 in the Prussian Academy edition.
7. H. Tabuchi and C. Jensen, "Yet Another Airbag Recall Will Affect Five Million," *New York Times*, February 4, 2016.
8. UN General Assembly, *Universal Declaration of Human Rights* (New York: UN General Assembly 1948), article 14.
9. Robert Nozick, *Anarchy, State, and Utopia* (New York: Basic Books, 1974), 175.
10. Judith Jarvis Thomson, "The Trolley Problem," *Yale Law Journal* 94, no. 6 (1985): 1395.

PART

III

Six Key Issues in Engineering Ethics

CHAPTER 7

Whistle-blowing: Should You Ever Break with Protocol?

In the previous chapters, we introduced some concepts and ethical theories that are helpful for understanding real-world problems faced by engineers in their professional careers. In this and the next five chapters, we will discuss some of the most important issues in engineering ethics in greater detail, with the aim of showing how the concepts and theories we have learned can be put to work.

The first issue is whistle-blowing. Hardly any workplace is populated entirely by employees who always follow the law and never perform any morally questionable actions. If you learn that your organization is complicit in serious moral or legal wrongdoing, and you are unable to resolve this issue by talking to your supervisor or by using other established channels of communication, then your only possibility to stop the wrongdoing might be to become a whistle-blower. A whistle-blower is, roughly put, someone who breaks with protocol and bypasses the ordinary chain of command by, for example, contacting the press (external whistle-blowing) or the supervisor's supervisor (internal whistle-blowing) to reveal serious moral or legal wrongdoing.

The origin of the whistle-blower metaphor is unclear. Some say it alludes to a train sounding a whistle to warn people to get off the track. Others believe it refers to a police officer blowing a whistle to stop a crime; or a referee blowing a whistle when a foul is committed in soccer, basketball, or some other sport. Irrespective of which explanation is closest to the truth, it is worth keeping in mind that whistle-blowing is often associated with huge personal risks. Whistle-blowers sometime lose their jobs or get demoted. In some countries, including the United States, there are laws designed to protect whistle-blowers, but not every instance of whistle-blowing is protected by the law and the law does not always work as intended. The Snowden case, which is a highly controversial example of whistle-blowing that tends to cause a lot of disagreement, illustrates this point well. (See Case 7-1 on the next page.)

CASE 7-1

Edward Snowden and the NSA

Edward Snowden worked for fifteen months at the National Security Agency's regional operations center in Hawaii, which monitors electronic communication in China and North Korea. Snowden was employed by the consultant firm Booz Allen Hamilton, not directly by the NSA. During his fifteen-month stint at the NSA, he became increasingly aware of what he believed to be immoral and illegal practices. Among other things, he discovered that the NSA routinely monitored millions of phone calls and emails sent by lawful US citizens. According to Snowden's own account of the events that followed, he first raised his concerns with his supervisor, but to no avail. His supervisor told him to do his job and not worry about ethical and legal issues.

Frustrated with his supervisor's unwillingness to take action, Snowden decided to do what he could to change the system. On May 20, 2013, he flew from Hawaii to Hong Kong with thousands of classified documents stored on his hard drive, which he had secretly downloaded from the NSA's servers over a period of several weeks. In Hong Kong, Snowden met up with journalists from the *Washington Post* and the *Guardian*. He gave them

access to the classified information and permission to publish whatever they deemed relevant. According to his own account, he emphasized that not all documents might be suitable for publication because of the risks this could impose on NSA employees and others. In the months that followed, the leaked NSA documents resulted in numerous front pages in dozens of highly respected newspapers around the world. In addition to monitoring millions of emails and phone calls, it was revealed that US government agencies had been spying on leaders in several allied states, including German chancellor Angela Merkel. Her cell phone had been tapped for years.

On June 21, 2013, the US Department of Justice initiated an investigation into the theft of government property and two counts of violating the US Espionage Act of 1917. Two days later, Snowden left Hong Kong with the intention of traveling to Cuba via Moscow. There is no evidence that he took any sensitive information with him to Russia, but when he arrived in Moscow, he was told that his passport had been canceled by the US State Department. As of 2018, he is residing in Russia (see Figure 7.1).

Figure 7.1
Ceremony for the conferment of the Carl von Ossietzky Medal 2014 to Edward Snowden, awarded via Google Hangout. (Carl von Ossietzky was a German pacifist and recipient of the Nobel Peace Prize.)
Source: The photo is in the public domain.

The NSA later admitted that many of Snowden's allegations were true. New legislation was introduced to ensure that the government changed its practices. According to former Attorney General Eric Holden, Snowden "performed a public service by raising the debate that we engaged in and by the changes that we made."[1] Former White House National Security Staff Director Timothy Edgar wrote in September 2016 that "Snowden forced the NSA to become more transparent, more accountable, more protective of privacy. . . . For that the U.S. government has reason to say, 'Thank you, Edward Snowden.'"[2]

Despite all this, there is no consensus on whether Snowden is a traitor or a hero. If he were to return home, he would probably have to spend many years in prison.

Discussion question: Would it be morally wrong to punish Edward Snowden for violating the Espionage Act even if it is true, as Timothy Edgar argues, that this "forced the NSA to become more transparent, more accountable, more protective of privacy"?

THE DEFINITION OF WHISTLE-BLOWING

Before discussing whether whistle-blowing is ever morally permissible, it is helpful to define the concept. The following definition captures the most important aspects:

> A *whistle-blower* is someone who passes along information about what she or he justifiably believes to be serious moral or legal wrongdoing in an organization of which the person is a member to an internal or external party she or he is not authorized to contact with the intention to stop this wrongdoing.

The central elements of this definition are *information, justified belief, serious wrongdoing*, and the *intention* with which the information is passed along. Let us discuss each element in turn.

Note that whistle-blowers *merely* pass along information to others. Whistle-blowers do not actively stop the wrongdoing. The fictional agent James Bond is, for instance, not a whistle-blower. He actively tries to catch "the bad guys" himself; he does not solve problems by merely passing on information to others. By definition, whistle-blowers inform others and let them take appropriate action. Edward Snowden met with a group of journalists in Hong Kong, who informed the public about what was going on, which in turn triggered the authorities to act.

The seriousness of the wrongdoing is also crucial to the definition. Imagine, for instance, that you learn that your supervisor has done something that is wrong, but not *very* wrong. Perhaps she booked a room in four-star hotel although she was authorized to stay only in a three-star hotel. This wrongdoing is, it seems, so trivial that you would not qualify as whistle-blower if you were to tell the press, or the chairman of the board.

It should also be stressed that the whistle-blower must be *justified in believing* that the information he or she passes on is correct. It is not sufficient that the whistle-blower *merely* believes that someone in the organization is guilty of wrongdoing. Imagine, for example, that you work in the White House in Washington, DC. Last time you met the president in person, you thought he looked a bit tired and therefore came to believe that he is addicted to tranquilizers. If you have no further evidence for your belief but nevertheless decide to contact the press, you are merely passing on gossip. This would not qualify as whistle-blowing because you have so little warrant for your belief.

Epistemologists disagree on what counts as warrant, but a common proposal is that the belief must be produced by a reliable method.

That said, it is important to keep in mind that information provided by a whistle-blower may later turn out to be false. Whistle-blowers are not infallible epistemic agents. Although it is desirable to be right, being so is not an essential element of whistle-blowing. Strictly speaking, all that is required is that the information the whistle-blower provides is sufficiently reliable and specific to warrant the recipient's attention.

The recipient can be either an external or internal party. Whistle-blowers we read about in newspapers are often external whistle-blowers who contact journalists; but it may sometimes be at least as effective, and more appropriate, to contact the relevant government authorities. Unlike the press, they have the legal powers required for investigating and pressing charges against wrongdoers.

For reasons that are easy to understand, we seldom hear much about internal whistle-blowers. Many companies and organizations have hotlines employees can call if they become aware of serious wrongdoing; and in some cases, employees are even permitted to not reveal their name or identity to the employer. To what extent such hotlines for whistle-blowers actually work as intended is hard to tell. Because the sensitive nature of this matter, we typically have to trust the information provided by the organizations, at the same time as it may be in their best interest to not reveal the information they receive.

Another distinction that may sometimes be relevant is that between open and anonymous whistle-blowing. Open whistle-blowing refers to cases in which the informant reveals his or her identity to the recipient of the information (but sometimes not to the public), whereas anonymous whistle-blowing refers to cases in which this condition is not fulfilled. Whistle-blowers sometimes have good reasons for conceding their identity; but if the information is difficult to verify, it may be advantageous to let the recipient know the whistle-blower's identity. Information that comes from a known and verifiable source is more likely to be given serious attention, although that of course increases the risk of retaliation.

It is also helpful to say a few words about the whistle-blower's intention or motivation for blowing the whistle. Ideally, we would expect the whistle-blower's primary motivation to be to stop the wrongdoing (this is often called "preventive" whistle-blowing), not to punish his or her enemies in the organization ("retributive" whistle-blowing). That said, it is not inconceivable that some whistle-blowers are also motivated by revenge. If you have first tried to contact your supervisor about what you have good reason to believe is some serious moral or legal wrongdoing, and this did not lead to any change, then it is not inconceivable that this will lead to tensions between you and your supervisor. Therefore, at the time when you decide to contact others, it is not unlikely that retribution is part of your motivation. This is understandable. However, if your *primary* intention is to harm those whom you consider to be your enemies in your organization, you are not a whistle-blower.

WHEN IS WHISTLE-BLOWING MORALLY PERMISSIBLE?

The moral debate over whistle-blowing is complex. Some scholars are concerned that whistle-blowing is, by definition, disloyal, though others question that idea.[3] By passing on information to someone the person is not authorized to contact, the whistle-blower

CASE 7-2

Was It Ethical to Report the Employer to a Hotline? (BER Case 97-12)

Could it ever be unethical to "blow the whistle," provided that the information you have is factually correct and concerns serious wrongdoing? The NSPE Board of Ethical Review (BER) has emphasized that engineers must first give those in charge of the organization an opportunity to correct their mistake, as illustrated by the following case:

> Engineer A is employed by SPQ Engineering, an engineering firm in private practice involved in the design of bridges and other structures. As part of its services, SPQ Engineering uses a CAD software design product under a licensing agreement with a vendor. Although under the terms of the licensing agreement, SPQ Engineering is not permitted to use the software at more than one workstation without paying a higher licensing fee, SPQ Engineering ignores this restriction and uses the software at a number of employee workstations. Engineer A becomes aware of this practice and calls a "hotline" publicized in a technical publication and reports his employer's activities. Was it ethical for Engineer A to report his employer's apparent violation of the licensing agreement on the "hotline" without first discussing his concerns with his employer?

The Board's conclusion was that it was unethical for Engineer A to report what he believed to be a violation of the employer's licensing agreement to the hotline without first giving the employer an opportunity to correct the alleged mistake. The Board notes that although SPQ Engineering did not respect the law, there was no immediate danger to public health or safety. The Board also points out that engineers have a duty to be loyal to their employers and clients. Therefore, Engineer A should first contact his employer. If that effort falls flat, it would be appropriate to report the violation of the licensing agreement on the "hotline." However, if there had been an *immediate* danger to public health or safety, then it would have been permissible to call the hotline right away.

Discussion question: Which of the ethical theories discussed in chapters 5 and 6 fits best with BER's analysis of this case?

---

violates a duty to be loyal to the organization of which she or he is a member. How could this ever be morally acceptable?

We shall consider three justifications for whistle-blowing. The first focuses on the *prevention of harm*: by blowing the whistle, the whistle-blower initiates a causal process that leads to the prevention of serious harm. This view resonates well with utilitarian theories. The second justification focuses on the distribution of guilt. Michael Davis, who has defended this view at length, argues that "[Whistle-blowers] are generally deeply involved in the activity they reveal. This involvement suggests that we might better understand what justifies (most) whistleblowing if we understand the whistle-blower's obligation to derive from *complicity* in wrongdoing rather than from the ability to prevent harm."[4]

The third justification holds that whistle-blowing is justified whenever that increases people's *autonomy*. On this view, Snowden's actions were morally right because the attention he got by the press ultimately made all of us more autonomous. The NSA no longer monitors emails and phone calls in the way it did in the past, meaning that we are now free to write and say what we want. For obvious reasons, the autonomy argument appeals to Kantians.

Richard T. DeGeorge defends the harm-preventing view. He claims that whistle-blowing is morally *permissible* if the first three criteria in the following list are satisfied and *obligatory* if and only if all five are satisfied:

1. A practice or product does or will cause serious harm to individuals or society at large.
2. The charge of wrongdoing has been brought to the attention of immediate superiors.
3. No appropriate action has been taken to remedy the wrongdoing.
4. There is documentation of the potentially harmful practice or defect.
5. There is good reason to believe public disclosure will avoid the present or prevent similar future wrongdoing.[5]

A possible objection to criterion 1 is that this is a very strong requirement. Would it not be sufficient to require that whistle-blowers have *good reason to believe* that a product does or will cause serious harm? Whether the product or practice *actually* causes any harm may not be a decisive moral factor. Criterion 5 is also problematic. It entails that whistle-blowing is obligatory *only if* there is some good reason to believe that public disclosure would stop or prevent the practice causing harm. However, claims about the effects of a complex action can seldom be substantiated before the action has been performed. Some would say that we almost *never* have good reasons to believe that an act of whistle-blowing will stop or prevent the practice causing harm before we have actually blown the whistle. The outcome depends on what is in the news the day the story breaks, and the public reaction to a news story is almost impossible to predict. Therefore, it seems that according to criterion 5, whistle-blowing would almost never be obligatory, which seems wrong. A more nuanced view could be that whistle-blowing is obligatory as long as it is not reasonable to expect that public disclosure would *not* have any effect.

These objections do not show that there is anything fundamentally wrong with the harm-preventing view. All that follows is that DeGeorge's version may not be entirely correct. Utilitarians would certainly agree that *some* version of the harm-preventing view, specified in the appropriate way, is correct.

Michael Davis summarizes his complicity-avoiding analysis of whistle-blowing in six criteria. In his view, you are obliged to blow the whistle if and only if

1. What you will reveal derives from your work for an organization;
2. You are a voluntary member of that organization;
3. You believe that the organization, though legitimate, is engaged in serious moral wrongdoing;
4. You believe that your work for that organization will contribute (more or less directly) to the wrong if (but *not* only if) you do not publicly reveal what you know;
5. You are justified in beliefs 3 and 4; and
6. Beliefs 3 and 4 are true.[6]

The central criterion of this theory is 4. This criterion entails that if you do not contribute to the wrongdoing yourself, then you have no obligation to become a whistle-blower even if all other criteria are met. But is this true? Imagine that you know for sure that a colleague working in a separate division of your company is deeply engaged in some serious moral wrongdoing. Your colleague is working on a very different type of product, and you are in no way "complicit" in what is happening in the other division. It would be odd to say that you are free to ignore this wrongdoing just because you are not complicit in it. We would not tolerate such passive behavior in other situations. If you learn that your friend has robbed a bank, you must report that to the police, even though you were not complicit in the crime yourself. This objection to the complicity-avoiding view is, arguably, more fundamental than the objection to the harm-preventing view raised previously.

The autonomy-based justification for whistle-blowing can be obtained from the complicity-avoiding view by replacing 4 with the following criterion:

> 4' You believe that if you do not publicly reveal what you know, then this will diminish people's autonomy by, for example, not making it possible to make informed decisions about what risks to accept.

The notion of autonomy in 4' should be understood broadly. In the Snowden case, the autonomy of ordinary citizens was violated in the sense that they were misled to believe that their emails were private, although they were not. If they had known what the NSA was up to, that might have altered their communicative behavior, meaning that they were not able to make informed decisions about what to write in emails and what not to write.

Which of these three justifications of whistle-blowing we think is best will depend on which ethical theory we accept. As mentioned earlier, utilitarians will feel inclined to accept the harm-preventing view, at least under some reasonable assumptions about the consequences of particular instances of whistle-blowing. Kantians are more likely to accept the complicity-avoiding theory or the autonomy-based justification. Virtue ethicists would arguably feel attracted by the complicity-avoiding view, while natural rights thinkers would reject the idea that one sometimes has a positive obligation to become a whistle-blower.

CASE 7-3

An Exemplary Whistle-blower?

Evan Vokes graduated from the University of Alberta with a degree in metallurgic engineering. About a decade after graduation, he found himself working for TransCanada, one of the largest developers and operators of pipelines in North America. While working there, Mr. Vokes began to notice that Trans-Canada did not always build their pipelines the way regulators demanded they should be built. Critical welds were not welded in the right way, and some pipelines had minor dents. He believed this could cause ruptures and worried what would happen if a pipeline were to explode in the vicinity of a big city.

Mr. Vokes first raised his concerns internally, but was told not to worry. One of his managers,

David Taylor, wrote in an email that he should "accept where we are and become aligned with where we are going as a company."[7] Disappointed with this answer, Mr. Vokes then decided to go all the way to the top. He met with TransCanada's Vice President for Operations and wrote a letter to the company's CEO, Mr. Russ Girling. His efforts led nowhere. In November 2011, Mr. Vokes went on stress leave.

While on leave, Mr. Vokes continued to worry what could happen if a pipeline were to explode. In March 2012, he finally decided to go outside the normal chain of command by meeting up with regulators from the National Energy Board (NEB) in Canada.

(Continued)

The NEB encouraged him to file a formal complaint, which he did on May 1. In a twenty-eight-page letter, Mr. Vokes detailed his concerns and gave examples of what he claimed to be violations of NEB rules by TransCanada. He also included photos from pipelines he had worked on. A week later, on May 8, TransCanada fired Mr. Vokes without cause.

A few months later, on October 12, the NEB issued a public letter to TransCanada in which it acknowledged that Mr. Vokes's complaints were valid: "many of the allegations of regulatory non-compliance identified by the complainant were verified by Trans-Canada's internal audit."[8] The NEB also wrote that the agency was "concerned by TransCanada's non-compliance with NEB regulations, as well as its own internal management systems and procedures." Around this time, a second employee at TransCanada contacted the NEB with similar concerns. Of the sixteen allegations filed by the second whistle-blower, NEB determined that six could be partially substantiated. However, the NEB concluded that none of them posed any "immediate" threat to the public.

On January 25, 2014, a natural gas pipeline owned and operated by TransCanada exploded about thirty miles south of Winnipeg. The flames were estimated to be 200–300 m high. Because the explosion occurred in a sparsely populated area, no one was killed or injured; but thousands were affected by a prolonged gas outage in the midst of winter. The NEB investigation revealed that the explosion had been caused by a pre-existing crack in the pipeline, which had gone unnoticed ever since it was built decades earlier. As the temperature dropped to –30 °C, the metal in the pipeline contracted, and the crack triggered a catastrophic rupture. This was in line with what Mr. Vokes had warned could happen.

In October 2014, Mr. Vokes received a "Special Award for Whistle-blowing" from the Council of Canadians. However, he has not been able to keep a permanent position since he was fired by Trans-Canada. The personal consequences for Mr. Vokes have not been favorable.

Discussion question: The TransCanada case fits well with DeGeorge's harm-preventing view: Mr. Vokes had good reason to believe that public disclosure would stop TransCanada's wrongdoing, so it seems that on this account, his actions were fully justified. It is less clear whether advocates of the complicity-avoiding and autonomy-based accounts could reach the same conclusion. Did Mr. Vokes do the right thing according to those accounts of whistle-blowing?

ADVICE TO WHISTLE-BLOWERS

It is not uncommon for whistle-blowers to be dismissed or punished by their employers. Many organizations *say* they welcome and respect whistle-blowers, but they do not always do so when push comes to shove. As I write this, Edward Snowden is living in exile in Russia, and Evan Vokes has so far not been able to find another job in the oil and gas industry. In light of these and other similar examples, it might be prudent to think twice before you decide to become a whistle-blower.

A first piece of advice to any potential whistle-blower is to book a meeting with a lawyer. The legal aspects of whistle-blowing vary from country to country. In the United States, whistle-blowers are protected by the Whistleblower Protection Act of 1989 (see Figure 7.2). This federal law stipulates that employees working for the government should not face retaliation if they report misconduct within their organization. However, despite this legal protection, retaliation is common. In 2015, the US Department of Labor received 3,337 reports of retaliation against whistle-blowers, of which 843 were settled with a positive outcome for the complainant.

A second piece of advice is to make sure that you can *prove* your claims to others. It will typically not be sufficient to just tell someone what you believe or have heard. You need documents or other hard facts that support your claims. Emails might work, but it is important that people who read them without having any background information come to the same conclusion as you. Ideally, the evidence should be so convincing that

Figure 7.2
Whistle-blowers are protected by the Whistleblower Protection Act. The picture shows the US Supreme Court in Washington, DC.   Source: iStock by Getty Images.

no reasonable person can ignore or deny the whistle-blower's claims. It may not be sufficient to rely entirely on witness reports from others.

A third and final piece of advice is to try to be as kind as you can to your organization. You may wish to offer your supervisor more than one chance to correct his or her mistake. A single email or meeting may not be sufficient. Just because your immediate supervisor turns you down in a meeting, it does not follow that it is prudent to contact the press or the CEO right away. Perhaps you can approach your supervisor together with a colleague who could help you make your point, or perhaps you could contact someone else in your organization who could talk to your supervisor. Becoming a whistle-blower should be your last resort if nothing else works out. Most supervisors in most organizations are reasonable people who are able and willing to listen to their employees. If you have to deal with someone who isn't, it may be wise to document your attempt(s) to solve the problem internally before you use other methods. To minimize the risk of retaliation, you need to convince your organization that what you did was right and that you have given them plenty of opportunities to correct their mistake.

REVIEW QUESTIONS

1. What did Edward Snowden do when working for the NSA and why were his actions so controversial?
2. Is Edward Snowden a whistle-blower?
3. Was Roger Boisjoly (see chapter 2) a whistle-blower?
4. Propose, by using your own words, a precise definition of whistle-blowing.

5. Apply DeGeorge's harm-preventing theory of whistle-blowing to the Edward Snowden case. Was Snowden obliged to blow the whistle according to DeGeorge's theory?

6. Apply Davis's complicity-avoiding theory of whistle-blowing to the Edward Snowden case. Was Davis obliged to blow the whistle according to DeGeorge's theory?

7. To what extent, if any, do the theories proposed by DeGeorge and Davis appeal to Kantian and utilitarian moral considerations or intuitions?

REFERENCES AND FURTHER READINGS

Burrough, B., S. Ellison, and S. Andrews, April 23, 2014. "The Snowden Saga: A Shadowland of Secrets and Light." *Vanity Fair.*

Davis, M. 1996. "Some Paradoxes of Whistleblowing." *Business & Professional Ethics Journal* 15, no. 1: 3–19.

De George, R. T. 1981. "Ethical Responsibilities of Engineers in Large Organizations: The Pinto Case." *Business & Professional Ethics Journal* 1, no. 1: 1–14.

De Sousa, M. April 13, 2014. "Transcanada Corp. Dismissive of Employees' Concerns about Pipeline Safety, Records Reveal." *The Star.* Toronto, Canada.

De Sousa, M. March 18, 2016. "They Told Me to Take Money and Run, Says Pipeline Whistle-blower." *National Observer.*

Gellmann, B. December 23, 2013. "Edward Snowden, after Months of NSA Revelations, Says His Mission's Accomplished." *Washington Post.*

Sawa, C. S. T., and J. Loeiro. October 18, 2012. "Investigation Follows Revelations from Whistle-blower." *CBC News.* See http://www.actionnews.ca/newstempcb.php?article=/news/canada/regulator-probing-safety-culture-at-transcanada-pipelines-1.1228488

Stieb, J. A. 2005. "Clearing Up the Egoist Difficulty with Loyalty." *Journal of Business Ethics* 63, no. 1: 75–87.

NOTES

1. Amnesty International and Human Rights Watch, published as an ad in the *New York Times*, September 21, 2016, p. A24.

2. Ibid.

3. Some scholars have questioned this view. See J. A. Stieb, "Clearing Up the Egoist Difficulty with Loyalty," *Journal of Business Ethics* 63, no. 1 (2005): 75–87.

4. Michael Davis, "Some Paradoxes of Whistleblowing," *Business & Professional Ethics Journal* 15, no. 1 (1996): 10.

5. Richard T. DeGeorge, "Ethical Responsibilities of Engineers in Large Organizations: The Pinto Case." *Business & Professional Ethics Journal* 1, no. 1 (1981): 6.

6. Davis, "Some paradoxes of Whistle-blowing," 11.

7. Vokes in De Sousa, M., "Transcanada Corp. Dismissive of Employees' Concerns about Pipeline Safety, Records Reveal," *The Star*, April 13, 2014. See https://www.thestar.com/news/canada/2014/04/13/transcanada_corp_dismissive_of_employees_concerns_about_pipeline_safety_records_reveal.html

8. C. S. T. Sawa and J. Loeiro, "Investigation Follows Revelations from Whistle-blower," CBC News, October 18, 2012. See http://www.actionnews.ca/newstempcb.php?article=/news/canada/regulator-probing-safety-culture-at-transcanada-pipelines-1.1228488

CHAPTER 8

Conflicts of Interest: When Is It Permissible to Influence the Actions of Others?

Imagine that you have just started a new job at the Federal Aviation Administration (FAA). Every now and then you meet with airline representatives to discuss technical aspects of maintenance procedures for aircraft engines. After a few months on the job, a small and relatively unknown airline invites you to its annual safety conference in Paris, France. The airline offers to cover international first-class tickets for you and your partner and three nights in a junior suite in the five-star Ritz Paris at Place Vendôme, just around the corner from the Louvre. In the invitation, it is made clear that the reason for inviting you is strictly professional and that the airline does not expect to receive any favor from you in return for the trip to Paris. Can you accept the invitation to attend the conference under these conditions?

According to US law, a bribe is "the offering, giving, receiving, or soliciting of something of value for the purpose of influencing the action of an official in the discharge of his or her public or legal duties."[1] A prosecutor seeking to get a suspect convicted for bribery must therefore demonstrate that there is a *direct connection* between the giving of the bribe and some specific past or future action performed in return for the bribe. For this reason, donations to political campaigns do not count as bribes in a legal sense. The connection between the donation and the future behavior of the politician is not sufficiently direct. (Surprisingly, this holds true even if the donation is very large and increases after each election as the candidate's voting pattern agrees more and more with the donor's viewpoints.) For the same reason, your luxury trip to Paris is not a bribe in a strict legal sense. The prosecutor would not be able to demonstrate a sufficiently direct connection between regulatory decisions taken by the FAA and your stay in a junior suite at the Ritz in Paris.

Despite this, it would nevertheless be *unethical* to travel to Paris under the circumstances just described (see Figure 8.1). This is because the trip triggers, at the very least, an apparent *conflict of interest*. No matter how hard you try to be objective and fair when

Figure 8.1

Is the person who paid for your first-class ticket trying to influence you?   Source: iStock by Getty Images.

you return to your job at the FAA, neutral observers can reasonably doubt that you in your role as regulator treat all airlines equally. It is not unreasonable to expect that you will be more generous to airlines that invite you to fancy conferences in luxury hotels. This holds true even if you *actually* manage to remain completely objective and neutral. Appearance counts.

The FAA and other government organizations recognize this distinction between the narrow legal definition of bribery and other, more wide-ranging ethical concerns related to the giving and acceptance of gifts and services. For instance, the FAA's Standards of Ethical Conduct explicitly "prohibit an employee from accepting gifts from sources seeking official action from, doing business with, or conducting activities regulated by the agency, or from sources with interests substantially affected by the employee's performance of duty."[2] For similar ethical reasons, employees are also prohibited from "holding stock or other securities in an airline, aircraft manufacturing company, or a supplier of components or parts to an airline or aircraft manufacturing company."[3]

This example serves as a reminder that ethical standards are often more demanding than the law. Although it is not criminal to attend luxurious conferences in foreign countries, it is sometimes unethical to do so. In this chapter, we will take a closer look at this and other ethical issues related to various types of conflicts of interest.

WHAT IS A CONFLICT OF INTEREST?

The NSPE code, and many other professional codes of ethics, stress the importance of disclosing or avoiding conflicts of interests. Consider the following NSPE Rule of Practice (Appendix A, II. Rules of Practice, 4a–c):

Engineers shall act for each employer or client as faithful agents or trustees.

a. Engineers shall disclose all known or potential conflicts of interest that could influence or appear to influence their judgment or the quality of their services.

b. Engineers shall not accept compensation, financial or otherwise, from more than one party for services on the same project, or for services pertaining to the same project, unless the circumstances are fully disclosed and agreed to by all interested parties.

c. Engineers shall not solicit or accept financial or other valuable consideration, directly or indirectly, from outside agents in connection with the work for which they are responsible.[4]

The mere fact that someone is facing a conflict of interest does not entail that the person is guilty of moral wrongdoing. This is a major difference compared to bribery, which appears to be wrong under almost all circumstances. A conflict of interest is morally problematic only as long as it is not properly *disclosed* or *resolved*. Consider, for instance, the married couple Alyssa and Bayo. Both of them work for North Texas Engineering Inc. Alyssa is Bayo's supervisor. She feels that Bayo deserves a promotion, but realizes that she is facing a serious conflict of interest: her assessment of Bayo may be influenced by personal considerations that are professionally irrelevant. The following general definition of a conflict of interest, proposed by Dennis F. Thompson, captures this worry well:

> A conflict of interest is a set of circumstances that creates a risk that professional judgment or actions regarding a primary interest will be unduly influenced by a secondary interest.[5]

As Bayo's supervisor, Alyssa's primary interest is to make a fair and objective evaluation of Bayo's work performance. However, because Alyssa is married to Bayo, there is a risk that her professional judgment will be unduly influenced by a secondary interest, such as the desire to improve the domestic atmosphere.

Note that the primary interest has to be related to some professional judgment or action. If different interests in your private life clash—perhaps you cannot figure out which of your many suitors to marry—you are not facing a conflict of interest. It is also worth stressing that it does not matter whether Alyssa's professional judgment is *actually* influenced by her secondary (domestic) interest. What matters is that there is a *risk* that her professional judgment is unduly influenced by a secondary interest. This point is connected to the distinctions between the following three types of conflicts of interest:

- Actual conflicts of interest
- Potential conflicts of interest
- Apparent conflicts of interest

You are facing an actual conflict of interest if your professional judgment regarding your primary interest is, or is at risk of being, unduly influenced by a secondary interest. If you are in a situation in which it can be reasonably foreseen that a conflict of interest is about to arise, then the conflict of interest is potential but not actual. Finally, the conflict of interest is apparent if a neutral bystander who is not fully informed of all relevant facts may reasonably suspect that your primary interest is at risk of being unduly influenced by a secondary interest.

In Alyssa's case, her conflict of interest is actual if she makes a professional judgment about her husband Bayo's work performance. Her conflict of interest is potential from the moment Bayo begins his job in Alyssa's department up to the point that Alyssa assesses his performance. During this time span, Alyssa's supervisor could avoid the emerging conflict of interest by asking someone else to assess Bayo. Finally, the conflict of interest is merely apparent if Bayo's colleagues have reason to believe Bayo will be assessed by his wife, but Alyssa has secretly asked her supervisor Charlie to assess Bayo because of the potential conflict of interest. In this case, there would be no actual or potential conflict of interest, but Bayo's colleagues could reasonably believe there is one.

As noted, the mere fact that a person is facing an actual, potential, or apparent conflict of interest does not mean that the individual is doing something unethical.

What matters is how that person deals with the conflict of interest. The worst thing you can do is to let a potential conflict of interest develop into an actual one without informing your supervisor or client. If you refrain from informing your supervisor or client, your behavior is likely to be unethical. What Alyssa should have done when Bayo started his job in Alyssa's department was to disclose this and discuss the situation with her supervisor. This holds true irrespective of whether Alyssa and Bayo are married or are just dating, or have some other type of close personal relationship.

CASE 8-1

Disclosing a Conflict of Interest (BER Case 93-6)

As mentioned in chapter 2, the NSPE Board of Ethical Review (BER) offers ethical guidance to its members on cases submitted for review. The following anonymized case was considered by the Board in 1996:

> Engineer A, a professional engineer in private practice, has been retained by Edgetown as town engineer. The Edgetown Planning Board has under review the approval of a project being proposed by ABC Development Enterprises. Engineer A is also being retained by ABC Development Enterprises on a separate project but that project is being constructed in Nearwood, a town in another part of the state. The Nearwood project is unrelated to the project under consideration by Edgetown. Engineer A is expected to offer his views in the capacity of town engineer to the feasibility of ABC's Edgetown project to the Edgetown Planning Board. Would it be ethical for Engineer A to develop and report his views on the feasibility study of ABC's Edgetown project to the Edgetown Planning Board?

In its analysis, the Board noted that two rules in the NSPE Code of Ethics are applicable to this case:

II.4.a. "Engineers shall disclose all known or potential conflicts of interest to their employers or clients by promptly informing them of any business association, interest, or other circumstances which could influence or appear to influence their judgement or the quality of their services."

II.4.d. "Engineers in public service as members, advisors, or employees of a governmental or quasi-governmental body or department shall not participate in decisions with respect to professional services solicited or provided by them or their organizations in private or public engineering practice."

The Board concluded that it would be ethically acceptable for Engineer A to report his views to the Edgetown Planning Board as long as he *discloses his relationship* with ABC Development Enterprises:

> Our position is based upon a belief that it is desirable to encourage small towns and municipalities to have access to competent engineering services at reasonable cost. . . . It is wholly unrealistic to interpret the Code to encourage engineers in private practice to perform in an advisory or other capacity to governmental bodies and at the same time bar them from performing any services as a consultant to that body. Such a narrow view would make engineers in private practice hesitant to accept advisory or other roles with public entities and deprive those entities of needed technical expertise.

The reasoning articulated by the Board in this particular case seems to be premised on some form of utilitarian reasoning. It is, ultimately, the *consequences* of the options available to the engineer that determined the Board's verdict. However, it is worth stressing that the Board is not in general committed to any particular ethical theory.

Discussion question: What is the point of *disclosing* an apparent, potential, or actual conflict of interest? Why does this eliminate or mitigate the moral concerns raised by the conflict of interest?

WHY CONFLICTS OF INTERESTS SHOULD ALMOST ALWAYS BE AVOIDED

Everyone agrees that engineers must disclose, and as far as possible avoid, conflicts of interest. However, to say that engineers must disclose and avoid conflicts of interests is like saying that water boils at 100 degrees Celsius. The statement is true under most circumstances, but exceptions are possible. Moreover, it remains to be seen *why* water boils at a certain temperature, or why conflicts of interest should be avoided.

Let us first discuss the possibility of exceptions. We know that at high altitudes, water boils below 100 °C, but could it ever be morally permissible for an engineer to conceal a conflict of interest? Unsurprisingly, we have to consider rather extreme situations to find convincing examples. During World War II, Oskar Schindler, a member of the Nazi party and owner of enamelware and ammunitions factories, saved more than 1,200 Jewish workers from being executed in concentration camps. He did so by

Figure 8.2

Oskar Schindler's grave in Jerusalem. By bribing Nazi officers toward the end of World War II, Schindler saved more than 1,200 Jews from the Holocaust.   Source: Public domain.

lying to his client, the Nazi army, about the workers' skills and talents. Schindler also bribed government officials on several occasions. Although Schindler was not an engineer, he had been enrolled at a technical school until he got expelled for forging a report card in 1924. We know that Schindler's professional actions were influenced by a secondary interest: his humanitarian desire to protect innocent human beings from being slaughtered by barbarian Nazis. Even though Schindler did not act as a "faithful agent or trustee" for his client, the army, and despite the fact that his professional actions were influenced by a secondary interest, it is obvious that Schindler did the morally right thing. It would be absurd to claim that Schindler should be criticized for not acting professionally. When Schindler died in 1974, the Israeli government bestowed him with the honor of being buried on Mount Zion in Jerusalem (see Figure 8.2). He is the only ex-member of the Nazi party awarded this honor.

Few modern-day engineers are likely to work under circumstances as extreme as Oskar Schindler's, but it is conceivable that modern-day engineers working abroad (perhaps in some nondemocratic country) may occasionally have to choose between adhering to their professional code of ethics and bribing a government official to achieve some greater humanitarian good. The take-home message of the Oskar Schindler example is that the prohibition against conflicts of interest is not a universal rule. It is a rule with exceptions. To understand the basis of this rule, it is helpful to consider how it can be derived from various ethical theories.

From a utilitarian point of view, conflicts of interests should be avoided because they tend to lead to bad consequences. As is the case with nearly all ordinary moral rules, rule utilitarians find it easier to justify this rule than act utilitarians. For rule utilitarians, there is no need to investigate the specific consequences of not avoiding every particular conflict of interest. As explained in chapter 5, rule utilitarians believe we should act in accordance with a set of rules that would lead to optimal consequences if they were to be accepted by an overwhelming majority in society. It seems obvious that the rule holding that engineers should avoid conflicts of interests satisfies this criterion, even if it is difficult to determine what the precise consequences would be on each and every occasion.

Act utilitarians object that the rule utilitarian analysis seems to entail that Oskar Schindler acted wrongly when he allowed professional actions to be influenced by his humanitarian desire to protect innocent human beings from being murdered. Rule utilitarians, of course, defend themselves against this objection by modifying the rule a little bit. On their view, a better rule could be to avoid conflicts of interests *unless one's secondary interest is a noble humanitarian one.* However, rather than trying to formulate some set of very complicated rules about which secondary interests are permitted and not, act utilitarians believe that conflicts of interests need to be avoided only when the overall consequences are suboptimal. The mere fact that it is often difficult to know what the consequences will be is irrelevant.

Kantians believe that the engineer's obligation to avoid conflicts of interests is an imperfect duty. If we ask ourselves if we could rationally will that the maxim "Engineers should not be required to disclose conflicts of interests" were to become a universal law governing our actions, then it seems clear that the answer is no. A world in which engineers don't disclose or avoid all conflicts of interests is certainly *conceivable* (so this is not a perfect duty), but we cannot *rationally will* that all of us were to live in such a world. A rational agent can will only to live in a world in which engineers are honest and inform their superiors and clients about conflicts of interests. Therefore, engineers have an imperfect duty to do so.

For virtue ethicists, the key issue concerns how responses to conflicts of interest manifest virtues such as justice, courage, and prudence. An engineer who conceals and does not avoid conflicts of interest treats his colleagues and clients unfairly and so fails to display the virtue of justice. In many cases, disclosing or avoiding conflicts of interest may require courage. Finally, one generally finds disclosing or avoiding conflicts of interest prudent because not doing so invites bad consequences for one's career.

Thinkers who believe that moral rights are social constructions can construct a client's right to be informed about a conflict of interest in roughly the same way as rule utilitarians: the acceptance of this right tends to lead to optimal consequences for society in the long run. For natural rights theorists, it is somewhat more difficult to explain why a client would have a right to be informed about a conflict of interest. In many cases, the explanation is that one of the clauses in the contract the engineer signed with his or her client stipulated that conflicts of interest should be disclosed or avoided.

INTERNATIONAL ENGINEERING

Attitudes to bribery and other ethical issues vary around the world. In many countries, it is widely believed that nearly every conflict of interest ought to be avoided and that bribery must, consequently, be banned. In other countries, bribery is a natural and ever-present element of everyday life. It is tolerated in much the same way we tolerate traffic jams on the highway: We don't like it, but we cannot figure out how to get rid of this everyday phenomenon.

Transparency International is an international organization specialized in fighting corruption. According to a recent survey, the world's least corrupt countries are Denmark, Finland, and Sweden. The United States is ranked 16, Russia is ranked 119, and Somalia and North Korea share the rank of 167.[6] Unsurprisingly, the fact that corruption is widely tolerated in many countries sometimes leads to trouble for engineers seeking to do business abroad. In countries at the lower half of the ranking, it might be virtually impossible to conduct business without paying large "fees" to cadres of middlemen. Should engineers ever tolerate this?

A possible answer is that we should accept the fact that not all countries are free of corruption, meaning that it is sometimes unavoidable, and therefore permissible, to pay whatever bribes are required for conducting business abroad. "While in Rome do as the Romans do," as the saying goes. Until a few decades ago, some professional codes of ethics contained explicit provisions that permitted engineers to follow local traditions and customs when conducting business abroad, even when that violated other parts of the code. However, in recent years, exceptions for foreign cultural norms have been widely abandoned. To some extent, this shift may have been influenced by the Foreign Corrupt Practices Act (FCPA) of 1977.

The FCPA makes it unlawful for US citizens and corporations to make payments to foreign officials to assist in obtaining or retaining business, even if such payments are legal in the country in which they are made. Consider the following example. Rex works for American Oil Inc. in Houston, Texas. On a business trip to Siberia, he finds out that the only way his company can obtain a permission to drill in Siberia is by donating a large amount to a "charity organization" in Switzerland, controlled by some local politician. Rex consults with Russian lawyers and discovers that it would not be illegal according to Russian law to make donations to the "charity organization" under the circumstances just described. A few weeks after American Oil Inc. makes a large donation to the Swiss charity, the company gets its permission to drill in Siberia.

In this case, neither Rex nor his employer did anything unlawful according to the laws of Russia. Despite this, Rex can still be convicted for bribery in the United States per the FCPA.

The FCPA recognizes a distinction between bribes and extortion payments. If you are forced to pay for something you have a prior *right* to, then you are the victim of extortion. This is not illegal. The Department of Justice (DOJ) explains that "a payment to an official to keep an oil rig from being dynamited should not be held to be made with the requisite corrupt purpose," meaning that this payment would not be illegal.[7] However, mere *economic* coercion does not count as extortion. The DOJ explains that "The defense that the payment was demanded on the part of a government official as a price for gaining entry into a market or to obtain a contract would not suffice since at some point the U.S. company would make a conscious decision whether or not to pay a bribe."[8]

Because the United States is such a large and economically powerful country, the FCPA has implications for business practices around the world. In a growing number of cases, the American DOJ has held foreign corporations accountable under the FCPA for transactions in a third country, because some of the transactions have been routed via American banks, or because one of the companies owns subsidiaries in the United States. If we believe corruption impedes economic growth—and threatens key societal values such as transparency, stability, and security—we should welcome this.

CASE 8-2

TeliaSonera in Uzbekistan

Founded in 1853, TeliaSonera is the largest and oldest telecommunications company in Sweden and Finland. At the beginning of the twenty-first century, TeliaSonera entered a series of emerging markets in Eastern Europe and Central Asia. The company currently has more than 200 million customers in eighteen countries. About 37 percent of the company is owned by the Swedish government.[9]

TeliaSonera entered the telecommunications market of Uzbekistan in 2007. Transparency International ranks Uzbekistan as one of the world's most corrupt countries, ranked 153rd out of the 167 countries on its list. Mr. Islam Karimov was the president of Uzbekistan from 1991, when the country declared independence from the Soviet Union, until his death in 2016. None of the elections held between 1991 and 2016 were free and democratic according international observers.

In 2012, a Swedish TV documentary showed that TeliaSonera paid $320 million in exchange for licenses and mobile phone frequencies in Uzbekistan to a company in Gibraltar owned by Gulnara Karimova, the daughter of Uzbekistan's dictator Islam Karimov. The Gibraltar-based company had no employees and no assets other than licenses and frequencies for conducting telecommunication operations in Uzbekistan (see Figure 8.3). TeliaSonera did not ask how the company had obtained the licenses and frequencies they so desperately needed. In an article in the British newspaper *The Guardian*, a group of US businessmen claimed in 2010 that "after [the businessmen] rejected Gulnara's offer to take a share in their Skytel mobile phone firm, 'the company's frequency has been jammed by an Uzbek government agency.'"[10]

Figure 8.3

The Telecom Tower in Tashkent, Uzbekistan. TeliaSonera paid $320 million in exchange for licenses and mobile phone frequencies in Uzbekistan to a company in Gibraltar owned by Gulnara Karimova, the daughter of Uzbekistan's dictator Islam Karimov.
Source: iStock by Getty Images.

According to leaked diplomatic cables, US diplomats in Uzbekistan reported back to Washington that Gulnara Karimova "bullied her way into gaining a slice of virtually every lucrative business" in Uzbekistan and is a "robber baron."[11]

TeliaSonera initially denied all allegations of wrongdoing. In their view, it was neither illegal nor unethical to acquire a small Gibraltar-based company for gaining access to telecom licenses and frequencies in Uzbekistan, regardless of the company's ties to Gulnara Karimova. A few months later, in 2013, a Swedish journalist received a phone call from an anonymous source in Moscow. The caller said he had some additional information that could further substantiate the allegations against TeliaSonera. The journalist spoke with the informant at a secret meeting in the Cadier Bar in the Grand Hotel in Stockholm. The documents appeared to show Gulnara Karimova's handwritten comments on various business transactions with TeliaSonera and other international companies in Uzbekistan. It turned out that TeliaSonera had been asked, and made, additional "donations" to a number of "charity organizations" in Uzbekistan.

John Davy, the Chief Financial Officer for TeliaSonera in Uzbekistan in 2008, confirmed in a filmed interview that all donations to "charity organizations" were bribes. He explained that every high-ranking government official in Uzbekistan runs a "charity organization" and that it is practically impossible to do business without making donations to them.

From time to time, TeliaSonera was asked to make substantial donations of about $100,000 to stay in business. If the "donations" came in late, the company's base stations were temporarily shut down by the government.

The CEO of TeliaSonera resigned in February 2013; and because the bribes had been routed from Sweden to Gibraltar via American banks, the US Department of Justice (DOJ) decided to take action. In 2016, the DOJ proposed that TeliaSonera should pay a fine of $1.6 billion to the US taxpayers. At the time of this writing, no final settlement has been reached. However, if the proposed fine is accepted by TeliaSonera, it will be one of the highest fines ever paid for violations of the Foreign Corrupt Practices Act (FCPA).

Supporters of the FCPA welcome the efforts by US authorities to reduce corruption worldwide, even when no US citizens are directly affected. Critics argue that fines paid for violations of FCPA are morally questionable. The victims of the TeliaSonera corruption scandal were not the taxpayers in the United States, but first and foremost the people of Uzbekistan. They, if anyone, deserve compensation.

Discussion question: Why do you think corrupt business practices are tolerated in some countries but not in others? What, if anything, can engineers do to reduce corruption?

REVIEW QUESTIONS

1. What is a bribe? Give a precise definition.
2. What is a conflict of interest? Give a precise definition.
3. How are conflicts of interest addressed in the NSPE Code of Ethics?
4. Explain the difference between actual, potential, and apparent conflicts of interests.
5. "To say that engineers must disclose and avoid conflicts of interests is like saying that water boils at 100 degrees Celsius." Explain this analogy and what, if anything, we can learn from it.
6. Oskar Schindler's professional actions during World War II were influenced by a secondary interest, namely, his humanitarian desire to protect innocent human beings from being murdered by barbarian Nazis. Did Schindler face an actual, potential, or apparent conflict of interest?
7. What is the Foreign Corrupt Practices Act?
8. Explain the distinction between bribes and extortion and how the Foreign Corrupt Practices Act addresses this distinction.
9. Explain the role played by "charity organizations" for TeliaSonera's operations in Uzbekistan.

REFERENCES AND FURTHER READINGS

Baruch, H. 1979. "The Foreign Corrupt Practices Act." *Harvard Business Review*, 57: 32–50.

Carson, Thomas L. 1985. "Bribery, Extortion, and 'The Foreign Corrupt Practices Act.'" *Philosophy & Public Affairs* 14, no. 1: 66–90.

Leigh, David. "WikiLeaks Cables: US Keeps Uzbekistan President Onside to Protect Supply Line." *Guardian*, December 12, 2010. Available online https://www.theguardian.com/world/2010/dec/12/wikileaks-us-conflict-over-uzbekistan

Milne, Richard. September 17, 2015. "TeliaSonera Set for Eurasia Exodus in Wake of Corruption Claims." *Financial Times*.

National Society of Professional Engineers. 2018. *Code of Ethics for Engineers.* Alexandria, VA. Available online: https://www.nspe.org/sites/default/files/resources/pdfs/Ethics/CodeofEthics/NSPECodeofEthicsforEngineers.pdf

Thompson, D. F. 1993. "Understanding Financial Conflicts of Interest." *New England Journal of Medicine* 329, no. 8: 573–573.

Transparency International. 2015. "Transparency International Corruption Perception Index 2015." www.transparency.org/cpi2015/

US Senate Report No. 114, 95th Congress, 1st Session (Washington, DC, 1977). Available online https://www.justice.gov/sites/default/files/criminal-fraud/legacy/2010/04/11/senaterpt-95-114.pdf

US Department of Transportation. 2011. "Supplemental Standards of Ethical Conduct for Employees of the Department of Transportation." Code of Federal Regulations 5 C.F.R. Part 6001. Washington, DC.

NOTES

1. Thomas L. Carson, "Bribery, Extortion, and 'The Foreign Corrupt Practices Act,'" *Philosophy & Public Affairs* 14, no. 1 (1985): 66–90, 71n11.

2. See www.transportation.gov/ethics (last accessed May 29, 2017).

3. US Department of Transportation, "Supplemental Standards of Ethical Conduct for Employees of the Department of Transportation," Code of Federal Regulations 5 C.F.R. Part 6001 (Washington DC, 2011), 845.

4. National Society of Professional Engineers, *Code of Ethics for Engineers* (Alexandria, VA, 2018), 1. Available online: https://www.nspe.org/resources/ethics/code-ethics

5. Dennis F. Thompson, "Understanding Financial Conflicts of Interest," *New England Journal of Medicine* 329, no. 8 (1993): 573.

6. Transparency International, "Transparency International Corruption Perceptions Index 2015," www.transparency.org/cpi2015/

7. US Senate Report No. 114, 95th Congress, 1st Session (1977). Available online https://www.justice.gov/sites/default/files/criminal-fraud/legacy/2010/04/11/senaterpt-95-114.pdf

8. Ibid.

9. Martin Peterson, the author of this text, owns stocks worth about $500 in TeliaSonera.

10. David Leigh, "WikiLeaks Cables: US Keeps Uzbekistan President Onside to Protect Supply Line," *Guardian*, December 12, 2010.

11. Ibid.

CHAPTER 9

Cost-benefit Analysis: Do the Ends Justify the Means?

Engineering projects sometimes fail because they are too costly or do not deliver sufficiently large benefits. A prominent example is the Concorde project, a supersonic jet airliner designed in the late 1960s. From a technical point of view, the Concorde was a masterpiece. Engineers working on the plane took great pride, and rightly so, in their technical achievement. However, the costs of the Concorde project turned out to significantly exceed its benefits. Seventeen airlines placed preliminary orders, but the operational and environmental costs turned out to be so high that only two airlines, British Airways and Air France, eventually bought the plane. Even though the Concorde cut the travel time between New York and London from six to three hours, the benefits of supersonic air travel did not compensate for the high fuel cost and noise pollution (supersonic booms). British Airways and Air France ceased all operations with the Concorde in 2003.

The aim of a cost-benefit analysis is to systematically assess the costs and benefits of all alternatives available to the decision maker before a decision is made. To facilitate such comparisons, all costs and benefits are routinely assigned monetary values. This includes items that it can be very difficult to evaluate in monetary terms, such as noise pollution (sonic booms) or the loss of human life in accidents. Critics of cost-benefit analysis object that the practice of assigning monetary values to non-market goods such as human life and pollution is unethical.

No matter whether this criticism is warranted, it is worth emphasizing that a cost-benefit analysis can never be morally neutral. The output of the analysis is a *ranking* of all options available to the decision maker; so in that sense, a cost-benefit analysis makes a claim about what ought or ought not to be done. Every claim about what someone ought or ought not do is a moral claim, at least in the absence of further caveats.

Ford Pinto and the Value of a Human Life

The Pinto (see Figure 9.1) was the first subcompact car developed by Ford for the North American market. It was designed to be an affordable domestic alternative to the cheap Japanese brands that had rapidly gained popularity in the United States in the 1970s. At that time, the industry average for developing new cars was forty-three months. The development team at Ford was therefore pleased to get the Pinto ready for production in just twenty-five months. Production began in 1971 and continued until 1980.

In 1973, Ford began to receive reports that several Pintos had caught fire in low-speed, rear-end collisions. Sometimes impact speeds as low as twenty miles per hour led to fatal fires. It turned out that the cause of the problem was the position and structural integrity of the fuel tank. Ford had become aware of the problem already in the development phase and made some minor changes to the fuel tank's position. It was now evident that this had not resolved the problem.

The same year the National Highway Traffic Safety Administration (NHTSA) circulated a proposal

Figure 9.1
Numerous Ford Pintos caught fire in low-speed rear-end collisions caused by a faulty design of the fuel tank. According to a cost-benefit analysis performed by Ford in 1973, the total benefit (for society, not Ford) of re-designing the fuel tank was worth $50 million; but the cost was estimated to be $137 million. Ford therefore decided to not redesign the fuel tank.
Source: Public domain.

for stricter fuel system regulations. The new regulations would, if adopted, prevent the fuel tank from exploding in low-speed, rear-end collisions. However, in a memo submitted by Ford to the NHTSA in 1973, Ford argued that the benefits of the proposed regulation would not outweigh its costs. According to Ford, a redesign of the fuel tank in accordance with the proposed regulation would eliminate about 180 burn deaths and 180 serious injuries per year, which Ford argued was worth $49.5 million for society. The cost for redesigning the fuel tank would be about $11 per vehicle. With a total of 12.5 million vehicles affected, the total cost would therefore be about $11 × 12.5 million = $137 million.

Here is Ford's cost-benefit analysis exactly as it is presented on page 6 in the memo sent to the NHTSA:

Benefits:

Savings	180 burn deaths, 180 serious injuries, 2100 burned vehicles
Unit cost	$200,000 per death, $67,000 per injury, $700 per vehicle
Total benefit	180 × ($200,000) + 180 × ($67,000) + 2100 × ($700) = $49.5 million

Costs:

Sales	11 million cars, 1.5 million light trucks
Unit cost	$11 per car, $11 per truck
Total cost	11,000,000 × ($11) + 1,500,000 × ($11) = $137 million

Ford's conclusion was that "The total benefit is ... just under $50 million, while the associated cost is $137 million. Thus, the cost is almost three times the benefits, even using a number of highly favorable benefit assumptions."[1] Unsurprisingly, Ford opposed the new regulation. They also decided not to change the original design of the fuel tank on the Pinto model, despite the growing number of reports about fatalities in low-speed, rear-end collisions.

Ford's cost-benefit analysis was based on information the company had obtained from the federal government. The $200,000 assigned to the loss of a statistical life was a figure supplied by the NHTSA, as was the $67,000 estimated cost to society of a serious injury. The figures included not just the direct costs of a fatality or injury caused by a rear-end collision (such as ambulances and fire trucks) but also the taxes the victim would have paid if he or she had been able to work.

When the press found out that a $11 modification per car had been rejected by Ford as being too expensive, even though this simple fix would have prevented hundreds of burn deaths and serious injuries, a heated debate broke out. Several critics argued they would have been more than happy to pay an extra $11 for helping to prevent 180 burn deaths and 180 serious injuries. The essence of their argument was that the cost of $11 per car did not outweigh the *moral* value of preventing 180 burn deaths and 180 serious injuries, no matter how many cars would be affected by this modification. In 1974, as many as 544,209 Pintos were sold; but due to the bad publicity, this number dropped to 223,763 in 1975. In 1978, a jury awarded $125 million in punitive damages and $2,841,000 in compensatory damages to a Mr. Grimshaw, who had been injured in one of the many Pinto fires. The judge later reduced the punitive damages to $3.5 million; but because of this legal setback, Ford decided to settle all similar cases out of court.

Discussion question: Is it morally wrong to assign a monetary value to the loss of a statistical life? Does the answer depend on whether you are a utilitarian, Kantian, or virtue ethicist?

THE VALUE OF A HUMAN LIFE

Ford's assumption that a statistical life was worth $200,000 was neither arbitrary nor groundless. This figure, calculated by the NHTSA, was meant to reflect the average person's contribution to the economy and the average cost for the emergency response. The value the NHTSA assigns to a statistical life has been continuously revised and updated since 1973. In 2016, the agency estimated a statistical life to be worth about $9.6 million.[2] This steep increase was not mainly due to inflation: $200,000 in 1973 would have been equal to about $1.1 million in 2016. However, the 2016 figure included a cost of more than $6 million for the loss of the deceased person's Quality Adjusted Life Years (QUALYs). The QUALYs are meant to reflect the value of the individual's life viewed from that person's own perspective. To put it crudely, a value of $6 million in QUALYs means that this amount of money is sufficient compensation for letting a fully healthy person die *n* years earlier than that person otherwise would have, where *n* is the number of years the average person has left to live.

Despite this numerical precision, it is sometimes objected that it is *unethical* to assign a monetary value to the loss of a human life. Even if we *can* put a value on a human life, it is far from clear that we *should* do that. One way to understand this objection is to maintain that every human life has *infinite* value. However, as attractive as this may sound, this would make it impossible to make reasonable decisions in cases in which more than one life is at stake. Imagine, for instance, that you can either rescue two people from a sinking ship or five hundred from another sinking ship. You cannot assist both groups. After some reflection, you decide to perform a cost-benefit analysis in which you assign infinite value to every person in both groups. In a strict mathematical sense, it then follows that saving two is as good as saving five hundred. This is because five hundred times infinity is equal to two times infinity. However, to claim that it would be permissible to rescue two people in a situation in which you could have rescued five hundred seems absurd.

Another way to understand the objection that it is unethical to assign a monetary value to the loss of a human life is to claim that a statistical life has no *precise* monetary value, or perhaps no monetary value *at all*. The idea that something has no precise value can be modelled by representing the value as an interval. Perhaps the monetary value of a statistical life can, say, be represented by interval between $6 million and $10 million.

Although this better reflects the uncertainties inherent in a cost-benefit analysis, it is a poor response to the claim that we should never assign monetary values to statistical lives. People who argue that we should refrain from this are not primarily concerned with the precision of monetary measures. Their objection is directed against the *very idea* of measuring something in monetary terms that ought not be measured in such ways. It does not matter if the value is set to a specific number or set of numbers. To measure the value of a human life in dollars is to compare apples with oranges. Money and a human life are *incomparable* entities, which cannot be (precisely or imprecisely) measured on a common scale.

Unfortunately, the claim about incomparability is hard to defend in contexts that require us to make a choice. In a TV debate about the Pinto case in 1977, economist and Nobel Prize laureate Milton Friedman rhetorically asked, "Suppose it would have cost 200 million dollars [per life saved] to fix the Pinto, what should Ford have done?" Friedman's point was that there must surely exist *some* amount of money that is worth more than a single human life. If so, dollars and human life are not *wholly* incomparable. As Friedman put it, "You cannot accept the situation that a million people should starve in order to provide one person with a car that is completely safe."[3] If society would have to spend a very large amount on preventing a single statistical death, then other valuable societal aims, such as eradicating poverty and starvation, would no longer be possible to pursue.

The upshot is that although difficult, we must sometimes assign a precise monetary value to a statistical life and make trade-offs between conflicting aims. We cannot make such trade-offs in a structured manner without comparing the alternatives available to us. It is true that these comparisons will sometimes be approximate and involve significant uncertainties. However, it is hardly an option to *refrain* from making comparisons in which human lives are balanced against other goods. Even if we do that, the choices we make will implicitly reveal how much money we assign to the loss of a statistical life. The relevant question to ask is, therefore, how these comparisons ought to be made. The key moral feature of cost-benefit analysis is that it includes *only* the consequences of our actions in the analysis. Critics of cost-benefit analysis, especially those who reject utilitarianism, believe other features of a choice situation may be morally important.

COST-BENEFIT ANALYSIS AND UTILITARIANISM

How intimate is the connection between cost-benefit analysis and utilitarianism? In the spreadsheet summarizing Ford's cost-benefit analysis of the Pinto's fuel tank (Case 9-1), the columns listing the monetary costs and benefits served as proxies for bad and good consequences. Utilitarians typically equate good and bad consequences with well-being and suffering (or happiness and pain, or the satisfaction of preferences), rather than with monetary costs and benefits. However, the structure of the two theories is the same. Both theories focus *exclusively* on the consequences of our actions, without paying any attention whatsoever to duties, rights, or virtues. Therefore, many of the moral objections that can be raised against utilitarianism can also be raised against the use of cost-benefit analysis.

Consider, for instance, Philippa Foot's objection, mentioned in chapter 5, that utilitarians must concede that doctors are sometimes morally obliged to kill healthy patients (against their will) if their organs could be transplanted to sufficiently many other patients in desperate need of transplant organs. If we perform a cost-benefit analysis of this choice, and make some reasonable assumptions about the total costs and benefits, we will arrive at the same conclusion as the utilitarian. The benefits of killing one innocent person to save five dying transplant patients outweigh the costs; but it seems absurd to conclude that we ought to kill healthy, innocent people to save a greater number of patients in need of transplant organs.

Although the utilitarian analysis of the transplant case differs in some minor respects from how this issue would be approached in a cost-benefit analysis (because the two theories focus on different features of the consequences), the weighing and balancing of good and bad consequences works in the same manner. A huge loss to a single individual can always be outweighed by sufficiently large benefits to sufficiently many others. It is just the aggregated goodness that matters. Each of us is worth nothing above and beyond our contribution to the aggregated sum total of valuable consequences for society as a whole. The Ford Pinto case is one of many illustrations of how this works in practice. As long as the numbers in the calculation were correct, it is indeed true that not fixing the problem with the fuel tank brought about the best consequences for society as a whole, although hundreds of innocent people had to pay with their lives.

Faced with this objection, advocates of cost-benefit analysis can choose between two types of responses. The first option is to bite the bullet and acknowledge the allegation that cost-benefit analysis is a form of utilitarianism. There are many things utilitarians could say in response to Foot's and many of the other objections, as noted in chapter 5; and despite all these objections, utilitarianism is still one of the most influential and widely accepted ethical theories.

The second option is to argue that the link between cost-benefit analysis and utilitarianism need not be as tight as many people seem to think. For those who are attracted by this option, there are several ways to soften the connection between the two theories. To start with, utilitarianism is a criterion of moral rightness. It tells us what *makes* right acts right, but it does not tell us how we ought to go about making real-world decisions. Cost-benefit analysis, on the other hand, is a tool for making real-world decisions. Because the two theories seek to answer different questions, it is thus not surprising that it is possible to accept one theory but reject the other. Note, for instance, that some advocates of the utilitarian criterion of moral rightness believe we should not use cost-benefit analysis for making important decisions about life and death. This is because the adoption of other decision-making procedures generates better consequences. Perhaps the best decision-making procedure is to follow some professional code of ethics, or common-sense intuitions. If you are a utilitarian, this all boils down to empirical questions about the consequences of adopting different procedures.

Many critics of utilitarianism agree that consequences matter in ethics, although they believe consequences are not *everything* that matters. Some even go as far as claiming that it is possible to assign values to nonutilitarian entities such as rights, duties, and virtues within a cost-benefit analysis. Philosopher Sven Ove Hansson writes that, "Deontological requirements can be included in a CBA [cost-benefit analysis], for instance by assigning negative weights to violations of prima facie rights and duties, and including them on the cost side in the analysis."[4]

A straightforward proposal for how to do this is proposed by Eyal Zamir and Barak Medina. Writing a few years after 9/11, they ask us to imagine a situation in which we have to determine "whether state officials should be authorized to shoot down an airplane suspected of being used as a weapon against human lives."[5] They propose that instead of calculating the costs and benefits of shooting down a high-jacked airplane with hundreds of passengers by subtracting the monetary costs from the monetary benefits, it would be more appropriate to apply a slightly more complex formula that explicitly assigns numbers to rights and duties. The formula they propose is the following:

$$(px - qy) - (K + K'y)$$

Zamir and Medina explain that "the variable x is the expected number of people who will be killed or severely wounded as a result of the terrorist attack; y is the expected number of innocent people killed or severely wounded as a result of the preventive action; p is the probability that the terrorists will successfully carry out their plan if no preventive action is taken; q is the probability that the preventive action will kill the passengers; and K and K' are two constants, jointly constituting a combined, additive and multiplier threshold."[6]

Let us take a closer look at the constants K and K'. In the somewhat technical language used by Zamir and Medina, these constants refer to "a combined, additive and multiplier threshold." What this means is that K and K' represent the moral value of violating rights or duties. K' represents a right or duty for which the value of the violation depends on how many people are affected; the negative value of violating K does not depend on how many individuals are affected.

Zamir and Medina's proposal for how to incorporate rights and duties in a cost-benefit analysis is, arguably, not perfect. But it shows that advocates of cost-benefit analysis need not be diehard utilitarians. It is indeed possible to include violations of rights and duties in a cost-benefit analysis. However, what makes Zamir and Medina's proposal difficult to apply to real-world cases is that it is no easy task to determine the precise values of K and K'. If we think it is hard to assign a monetary value to a human life, as noted earlier in this chapter, we will probably find it even more difficult to assign a precise numerical value to violations of rights and duties. (Note that if K or K' is infinite, then the overall value is minus infinity, no matter how large the finite benefits are.)

An alternative strategy for incorporating rights and duties in a cost-benefit analysis is to apply what Rosemary Lowry and I have characterized as *output* and *input* filters. An output filter is a device that filters out alternatives that violate people's rights or duties before the monetary costs and benefits of those options are determined. For instance, if we think consumers have an absolute right to be protected against cars that are likely to catch fire in low-speed, read-end collisions, then the option of not fixing the problem with the Pinto's fuel tank could be eliminated from the list of permissible actions by filtering out that alternative already before its monetary costs and benefits are determined. Input filters work in similar ways, except that they filter out certain pieces of information about the available options that should not be included in the analysis for moral reasons. Philosopher Sven Ove Hansson gives an example of what type of information it might be appropriate to filter out:

> Suppose that a CBA is made of a program against sexual assaults. The sufferings of the victims should clearly be included in the analysis. On the other hand, few would wish to include the perverse pleasures experienced by the perpetrators as a positive factor in the analysis. Generally speaking, we expect a credible CBA to exclude positive effects that depend on immoral behavior or preferences.[7]

If we believe that positive consequences that depend on immoral behavior or preferences count as rights violations, then these positive consequences could be ignored in the analysis by applying an input filter.

What all this shows is that there are several quite sophisticated ways in which advocates of cost-benefit analysis could render their preferred decision-making procedure compatible with fundamental nonutilitarian intuitions. To reject cost-benefit analysis for being "too utilitarian" seems to be a mistake.

CAN WE PUT A PRICE ON THE ENVIRONMENT AND HISTORIC ARTIFACTS?

Engineers regularly perform cost-benefit analyses of projects that may impact the natural environment or historic artifacts. However, unlike you and me, rocks and historic buildings have no desires or preferences, nor are they goods that can be bought on the market. There is, therefore, no direct way in which we can measure the economic impact of a lake, or historic building, in a cost-benefit analysis. Despite this, many of us would agree that we ought to preserve at least some historic buildings and attractive parts of nature just because of their intrinsic beauty.

While advocates of cost-benefit analysis admit that it can be difficult to assign monetary values to the natural environment and remains from the past, they do not believe it is impossible to do so. *Hedonic pricing* and the *travel cost method* are two commonly used methodologies for assigning monetary values to non-market goods.

To illustrate how hedonic pricing works, imagine two similar residential areas in the outskirts of some big city, one of which is located next to a clean, beautiful lake. The average price of property located next to the lake is likely to be higher than the price of houses located elsewhere. Many homebuyers are willing to pay more for houses located close to the lake. However, because houses are goods with many different attributes (they come in different sizes, styles, and overall conditions), the entire price difference cannot be attributed to the lake. The good news is that if the number of houses is large enough, and there is sufficient variation in the data set, we can apply statistical methods for calculating how much of the price difference should be attributed to each contributing factor, such as proximity to the lake. For instance, if a house located within ¼ mile of the lake costs on average $50,000 extra, and there are one hundred such houses, then the monetary value of the lake is at least $5 million.

The travel cost method also seeks to value goods in indirect terms. If, for instance, ten thousand people are willing to travel three hundred miles from a nearby city to visit a historic castle or beautiful forest, the value of the castle or the lake can be said to equate the total amount these visitors paid for traveling to the site and back again. So, by asking a representative group of visitors how much they paid for gas and lodging, we can figure out what the total value is by summing up these figures for visitors to the site. Naturally, this presupposes that the visitors' primary reason for embarking on the trip was to visit the site in question.

The advantages of hedonic pricing and the travel cost method is that they make it relatively easy to assign monetary values to items it would otherwise be difficult to put a price on. All we need to do to apply these methods is to collect and analyze some relatively easily accessible data. However, critics object that the methods do not adequately measure the value of, say, the lake itself. The methods just measure, indirectly, how much we are willing to pay for living next to the lake or for visiting it, but we cannot equate its true value with that very limited economic value.

Imagine, for instance, that you are working on a project that may affect a beautiful, clean lake in a remote part of New Mexico. There are no houses or other relevant market goods nearby, so the hedonic pricing method does not apply. Moreover, almost no one ever bothers to travel to the lake because of its remote location. So per the travel cost method, the value of the lake is close to $0. Does this mean the value of preserving the lake would be $0? Advocates of cost-benefit analysis are aware that this seems wrong. In response to that, they have introduced something they call *existence value*. The existence value of a clean, beautiful lake in a remote part of New Mexico that no one ever visits is determined by asking people how much they would, hypothetically, be willing to pay for preserving the lake as it is, for its own sake. The total value of the lake is then equal to the sum of all those values. So, if you are willing to pay a small amount and I am willing to pay roughly the same small amount, then the total value is the sum of all those small amounts.

The problem with measuring the existence value of something is that this measure reflects how much *we* are willing to pay for the mere existence of something. It seems possible that the value something has for its own sake may not be equal to how much the current population is willing to pay for that something for its own sake. If the object in question is a historic building, its existence value may very well increase in the future, if future generations are willing to pay more than we for preserving the building as it is. Moreover, it seems perfectly possible to imagine that the mere existence of something is valuable for its own sake, even if no one is willing to pay anything for its mere existence. To be more precise, there are at least two issues that we need to keep apart: (a) how much people are willing to pay for the existence of x, and (b) the value something actually has for its own sake. Consider, for instance, a house designed by the infamous architect Talentlessi. It is not entirely inconceivable that someone is willing to pay a huge amount, say $1 million, for preserving a building by Talentlessi for its own sake, despite the fact that its remote location entails that it has no market value. It does not follow from that that the building actually is valuable for its own sake. Perhaps the person who is willing to pay for preserving the building happens to be wrong and is willing to preserve something that is not worth preserving. Conversely, we can also imagine a situation in which no one is willing to pay for the mere existence of a remotely located historic building that deserves to be preserved because it was designed by Talentlessi's more talented cousin Peruzzi.

The upshot of all this seems to be that existing methods for valuing non-market goods do not fully capture all aspects of an object's value that might be relevant in a cost-benefit analysis. The existing methods make the scope of a cost-benefit analysis more limited than one might have wished.

REVIEW QUESTIONS

1. What is a cost-benefit analysis and what is the moral importance of such analyses?
2. Why was Ford's decision to not modify the design of the Pinto's fuel tank so controversial?
3. In 2015, the US NHTSA estimated a statistical life to be worth about $9.1 million. Is it unethical to assign a monetary value to the loss of a human life? If so, why?
4. Are defenders of cost-benefit analysis committed to utilitarianism?
5. How, if at all, could defenders of cost-benefit analysis take rights and duties and other nonutilitarian considerations into account?
6. What role could output and input filters play in a cost-benefit analysis?
7. What problem is hedonic pricing and the travel cost method supposed to solve?
8. How, if at all, can a cost-benefit analysis account for the "mere existence" of a beautiful clean lake in some remote part of the world? Is that account plausible?

REFERENCES AND FURTHER READINGS

Adler, M. D. 2011. *Well-being and Fair Distribution: Beyond Cost-benefit Analysis*. New York: Oxford University Press.

Birsch, D., and J. Fielder, eds. 1994. *The Ford Pinto Case: A Study in Applied Ethics, Business, and Technology*. Albany: State University of New York Press.

Boardman, A. E., D. H. Greenberg, A. R. Vining, and D. L. Weimer. 2017. *Cost-benefit Analysis: Concepts and Practice*. Cambridge University Press.

Grush, E. S., and C. S. Saundy. 1973. "Fatalities Associated with Crash Induced Fuel Leakage and Fires." Detroit, MI: Ford Environmental and Safety Engineering.

Hansson, S. O. 2007. "Philosophical Problems in Cost–benefit Analysis." *Economics and Philosophy*, 23, no. 2: 163–183.

Hubin, D. C. 1994. "The Moral Justification of Benefit/Cost Analysis." *Economics and Philosophy* 10, no. 2: 169–194.

Lowry, R. and M. Peterson. 2012. "Cost-benefit Analysis and Non-utilitarian Ethics." *Politics, Philosophy and Economics* 11, no. 3: 258–279.

US Department of Transportation. 2016. "Revised Departmental Guidance 2016: Treatment of the Value of Preventing Fatalities and Injuries in Preparing Economic Analyses." Washington, DC: US Department of Transportation.

Zamir, E., and B. Medina. 2008. "Law, Morality, and Economics: Integrating Moral Constraints with Economic Analysis of Law." *California Law Review* 96, no. 2: 323–391.

NOTES

1. E.S. Grush and C. S. Saundy, "Fatalities Associated with Crash Induced Fuel Leakage and Fires," (Detroit, MI: Ford Environmental and Safety Engineering, 1973), 8.

2. US Department of Transportation, "Revised Departmental Guidance 2016: Treatment of the Value of Preventing Fatalities and Injuries in Preparing Economic Analyses" (Washington, DC: US Department of Transportation, August 8, 2016), 2.

3. You can watch the video online: www.youtube.com/watch?v=lCUfGWNuD3c (last accessed May 29, 2017).

4. Sven Ove Hansson, "Philosophical Problems in Cost–benefit Analysis," *Economics and Philosophy* 23, no. 2 (2007): 164n2.

5. Eyal Zamir and Barak Medina, "Law, Morality, and Economics: Integrating Moral Constraints with Economic Analysis of Law," *California Law Review* 96, no. 2 (2008): 375.

6. Ibid. 376.

7. Hansson, "Philosophical Problems in Cost–benefit Analysis," 173–174.

CHAPTER 10

Risk and Uncertainty: How Safe Is Safe Enough?

About four hundred nuclear power plants are currently in operation worldwide, of which approximately one hundred are located in the United States. The accidents in Three Mile Island, 1979; Chernobyl, 1986; and Fukushima, 2011, show that large-scale radioactive processes a few miles from where people live and work will never be entirely risk-free. It is impossible to foresee, control, and prevent all ways in which technological systems can fail. However, the mere fact that accidents *can* occur does not, at least by itself, entail any definitive conclusion about the moral acceptability of these technological systems. (To argue that it does would violate Hume's Law discussed in chapter 4: a valid inference to a moral conclusion requires at least one moral premise.) The fundamental ethical question we should ask is, arguably, the following: Under what conditions is it morally acceptable for engineers to impose technological risks on others? This question demands a thoughtful answer.

Needless to say, the ethics of risk is not just a debate over the risks of man-made radioactive processes. Nearly *all* technologies expose us to some sort of risk, or are believed to do so by the public. To better understand why experts and laypeople sometimes disagree on the acceptability of technological risks, it is helpful to distinguish between *objective* and *subjective* risk. The objective risk of an activity depends on how likely some negative event is to actually occur. For instance, the objective risk of losing money if you bet on red when playing roulette in Las Vegas is approximately 50 percent. The *subjective* risk, also known as the *perceived* risk, depends on what you believe about the world. For instance, if you believe your favorite Las Vegas casino has tinkered with the roulette wheel, then the subjective risk could be much higher. Another example is the possible cancer risk associated with cell phones. In the 1990s, some laypeople believed that radiation from cell phones (i.e., ordinary electromagnetic radiation) could cause cancer. When scientists tested this hypothesis scientifically, it

Figure 10.1
DDT is a known carcinogen. It was banned by the US Environmental Protection Agency in 1972 but is still being used in some countries in Asia and Africa.   Source: iStock by Getty Images.

turned out that the subjective risk was much higher than the objective risk. We now know that ordinary electromagnetic radiation from cell phones does not cause cancer.[1]

As trivial as the distinction between objective and subjective risk may be, it is key to understanding why risk is such an important topic in engineering ethics. If an activity's subjective risk differs from its objective risk, then this often triggers tensions between engineers and the public. It is not sufficient to ensure that a product *is* safe. Engineers also need to pay attention to how new and existing technologies are *perceived*. Trying to calibrate an activity's subjective risk with its objective counterpart is neither irrational nor irrelevant. Sometimes the public is wrong, but engineers and scientists are not infallible either. Experts sometimes make incorrect risk assessments. For instance, when DDT (dichlorodiphenyltrichloroethane; see Figure 10.1) was introduced on a large scale in the agricultural sector in the 1950s, it was perceived as safe by experts as well as laypeople, meaning that its subjective risk was low. The inventor, Paul Muller, was awarded the Nobel Prize in Medicine for his work on DDT in 1948. However, in the 1960s, a group of scientists discovered that DDT is, as a matter of fact, extremely toxic. The objective risk turned out to be several magnitudes higher than the subjective risk. In 1972, the substance was banned by the US Environmental Protection Agency.

THE ENGINEERING DEFINITION OF RISK

Engineers often characterize risks in quantitative terms. According to what we may call the *engineering definition of risk*, the risk of some unwanted event *e* is the product of the probability that *e* will occur and the value of the harm caused by *e*, measured in whatever unit deemed appropriate. If, for instance, the probability is one in ten million that an airliner with 450 passengers will crash in any given year, then the risk is $0.0000001 \times 450 = 0.000045$.

The engineering definition of risk differs significantly from the everyday notion of risk. In everyday contexts, a risk is simply an unwanted event that may or may not occur. On this view, the risk is the event itself (the crash), not the number representing the product of probability and harm (0.000045). Most of us are aware that airplanes crash from time to time, but few of us know the precise probability of this type of event. If asked to specify a number that reflects both probability and harm, we could perhaps do so by guessing; but it is highly unlikely that the number we pick is close to the truth.

Because the engineering definition of risk provides decision makers with quantitative information, it may appear to be more objective and accurate than the everyday notion. However, it is not uncommon that we encounter situations in which it is very difficult, and even impossible, to apply the engineering definition of risk. The trichloroethylene case illustrates this point well.

CASE 10-1

How Toxic Is Trichloroethylene?

Trichloroethylene is a clear, nonflammable liquid commonly used as a solvent for a variety of organic materials. It was first introduced in the 1920s and widely used for industrial purposes until the 1970s. At that point, suspicions arose that trichloroethylene could be toxic, which led to several scientific studies.

In the 1990s, researchers at the US National Cancer Institute showed that trichloroethylene is carcinogenic in mice. Although many substances that are carcinogenic in mice are also carcinogenic in humans, the correlation is far from perfect. Some substances that cause cancer in mice do not cause cancer in humans. Therefore, these findings were not deemed sufficient for settling the scientific debate. Experts continued to disagree on whether trichloroethylene was a human carcinogen or not.

For ethical reasons, it was not possible to conduct experimental studies on humans. Scientists were not willing to risk someone's life just to find out if trichloroethylene is carcinogenic. The best information they could hope to obtain were data from epidemiological studies. However, the epidemiological studies turned out to be inconclusive. This sometimes happens when the effect one is looking for is small. The researchers therefore decided to compare and evaluate the results of many independent epidemiological studies.

After having reviewed all evidence available in 2011, the US National Toxicology Program's (NTP's) 12th Report on Carcinogens concluded that trichloroethylene can be "reasonably anticipated to be a human carcinogen."[2] Other organizations, such as the EPA, reached a more definitive view and concluded that trichloroethylene is a "known carcinogen." This was somewhat surprising because the conclusion of the EPA was based on exactly the same data as the 12th NTP report. Do we *know* that trichloroethylene is a carcinogen, or is it merely *reasonable to anticipate* that we will eventually discover this? If so, should trichloroethylene be banned?

Discussion question: How certain do we have to be for justifying a regulatory decision that has negative consequences for private business owners? Do we have to *know* that the product is, say, a carcinogen, or is it sufficient that it can be *reasonably anticipated* to be carcinogenic? Or would a *mere suspicion* be sufficient?

The scientific debate over trichloroethylene indicates that not all risks can, at least not initially, be adequately described in probabilistic models. Sometimes we simply have too little evidence for making a probabilistic risk assessment. Even if numerous scientists collaborate to research the properties of a chemical substance, they sometimes fail to reach consensus. All that can be concluded is that the substance "might," or "can," or "can be reasonably anticipated" to be carcinogenic. Unless we wish to base regulatory decisions on purely subjective probability assessments, the engineering definition of risk does not universally apply to all risks engineers encounter.

Having said all that, it is of course easy to imagine other situations in which the engineering definition of risk applies and yields reasonable conclusions. Suppose, for instance, that you have been asked to design a 56 ft. sailing boat. The boat must be light and fast but strong enough to withstand rough seas. It is widely known that containers and other large objects sometimes fall off of cargo ships. Because a fully loaded container is somewhat lighter than sea water, thousands of containers are currently floating around in the oceans. If a boat (see Figure 10.2) sails into a floating container, the hull will be severely damaged and the boat may sink.

The best way to make sure that boats hit by floating containers do not sink is to add an extra watertight bulkhead in the bow. Unfortunately, such bulkheads make the boat heavier and slower, and the bulkhead costs about $20,000 extra. Now ask yourself

Figure 10.2
The author's sailboat in Tilos, Greece. It has no watertight bulkhead in the bow.   Source: Author.

the following question: Should you, as the engineer responsible for the design of the boat, instruct the builder to spend an additional $20,000 on a watertight bulkhead as a precautionary measure?

If you were to base your decision on a *risk-benefit analysis*, you would first have to assess the probability of hitting a floating container. No one really knows the exact figure, and the number of floating containers per square mile depends on which sea you are sailing in. The number of floating containers in the Gulf of Mexico is lower than in the North Atlantic. Let's say that you, after having studied the literature, conclude that the probability of hitting a container is one in a million per one thousand miles sailed. The boat is expected to sail forty thousand miles in the coming forty years, meaning that the probability that it will *not* hit a container at least once is about $(1 - (1/10^6))^{40} = 0.99996$.

If the value of the boat is $500,000, your decision can be represented as a choice between two options with a total of four possible outcomes. The monetary value of each outcome depends on the cost of the bulkhead and on whether the boat sinks, which it will do only in case no bulkhead was added and the boat hits a container. See Table 10.1.

In a risk-benefit analysis, the *expected value* of each alternative is calculated by multiplying the probability of each outcome with the value of that outcome and then summing up the terms for each option. Therefore, the expected value of adding the watertight bulkhead is –$20,000. If you select this option, it does not matter whether the boat hits a container or not; and the expected value of not the adding the watertight bulkhead is $(1 - 0.99996) \times -\$500,000 = -\$19.90$. So, if we believe that an option's expected value should guide our actions in a quantitative risk-benefit analysis, it follows that it would be better to *not* install any watertight bulkhead because the cost of the bulkhead does not outweigh the benefit: –$19.90 > –$20,000.

Critics of a quantitative risk-benefit analysis point out that the numbers used in these analyses are often uncertain. How sure can you be that the probability of sailing into a container is one in a million per one thousand miles you sail, and not five, ten, or twenty times higher or lower? And what about the values we used in this example: Can *all* values that matter really be expressed in monetary terms?

Another problem is that a quantitative risk-benefit analysis tends to overlook features of a situation we intuitively consider to be relevant. Psychologists have, for instance, shown that people are more willing to accept *voluntarily* assumed risks (e.g., the risk of getting killed by an avalanche in a ski resort) than *involuntary* ones (e.g., the risk of getting killed in a terrorist attack). However, if the probability of both types of risks is the same, the risk-benefit principle will rank then as equally undesirable.

Some critics also argue that we should distinguish between risks with *reversible* and *irreversible* effects. If you lose a large amount of money in a Las Vegas casino, this effect is reversible; but if you do something that harms the environment, that will typically

Table 10.1 The expected value of each alternative in a risk-benefit analysis is obtained by multiplying the probability of each outcome with the value of that outcome and then sum up the terms for each option

	p(no container) = 0.99996	p(container) = 1 – 0.99996
Add extra bulkhead	–$20,000	–$20,000
No bulkhead	$0	–$500,000

be an irreversible effect. From a moral point of view, risks with irreversible effects seem more problematic than risks with reversible effects, all else being equal—partly because they leave future decision makers with less freedom of choice. In theory, a quantitative risk-benefit analysis could account for this difference by assigning a higher negative value to risks with irreversible effects.

Another distinction discussed in the literature concerns the properties of *short-term* and *long-term* risks. Short-term risks occur shortly after a decision is made (you lose your money in the casino immediately after you have placed your bet), whereas the negative consequence associated with a long-term risk may occur decades or centuries after the crucial decision was made. Because long-term effects tend to be more difficult to assess and evaluate than short-term ones, long-term risks are often more difficult to assess in a quantitative risk-benefit analysis.

THE PRECAUTIONARY PRINCIPLE

The precautionary principle is an influential alternative to traditional risk-benefit analysis, especially in Europe and other countries outside the United States. According to the precautionary principle, there is no need to determine what the probability of sailing into a floating container is, or what the probability of a nuclear meltdown might be. All we have to establish is that some *sufficiently bad* outcome *may* occur if no precautionary measures are taken.

The precautionary principle has been codified in numerous international documents on global warming, toxic waste disposal, and marine pollution. However, there is no such thing as *the* precautionary principle. There exists no single, universally accepted formulation of this principle. The closest we get to such a formulation is the one in the *Report of The United Nations Conference on Environment and Development*, commonly known as the Rio Declaration:

> Where there are threats of serious or irreversible damage, lack of full scientific certainty shall not be used as a reason for postponing cost-effective measures to prevent environmental degradation.[3]

Unlike other formulations of the precautionary principle, the Rio Declaration explicitly requires that measures taken in response to a threat have to be cost-effective. This appears to be a reasonable qualification. If two of more different measures address the same threat equally well, why not choose the most cost-effective one?

Note, however, that the notion of cost-effectiveness mentioned in the Rio Declaration does not tell us what to do in situations in which the precautionary principle and the cost-benefit principle clash. Would it, for instance, be morally right to implement a *less* effective measure against some threat if it is *more* cost-effective?

Another influential formulation of the precautionary principle is the Wingspread Statement:

> When an activity raises threats of harm to human health or the environment, precautionary measures should be taken even if some cause and effect relationships are not fully established scientifically.[4]

The Wingspread Statement omits cost-effectiveness but stresses that precautionary measures should be taken even before a threat has been "fully established scientifically."

Critics of the precautionary principle object that the history of science shows something perceived as a threat at one time may later be recognized as perfectly harmless. An individual's subjective risk is sometimes much higher or lower than the objective risk. When the first railroads were built in the nineteenth century, some people feared that traveling at a speed of twenty mph could make us sick. We now know that this worry was unwarranted and that it would have been a mistake to ban railroads in the nineteenth century because of this "threat."

Others object to the rigidity of the precautionary principle. It apparently demands that once we identify a certain level of risk, we must take preventive measures. Because virtually everything we do carries some risk of non-significant damage, the precautionary principle seems to require that every human activity be prohibited.

In response, note that most activities do not raise threats of harm to human health, at least not in any morally significant sense of "threat." Although the flapping of the wings of a distant butterfly may, in principle, cause the outbreak of World War III, we need not take precautionary measures against butterflies. This unlikely scenario could occur only in very distant possible worlds.

A third objection to the precautionary principle is that it is too imprecise to serve as a regulatory standard. Its logic is unclear and its key terms poorly defined. The best response here is to tighten the definition. Exactly what kind of precautionary measures should we take and under what circumstances? Consider the following suggestion for how we could try to render the precautionary principle more precise:

> A technological intervention is morally right only if reasonable precautionary measures are taken to safeguard against uncertain but non-negligible threats.

The key terms in this formulation are "reasonable . . . measure" and "non-negligible threat." Each term can yield various interpretations. It would be difficult to precisely define how costly a precautionary measure must be to be unreasonable. Similarly, we could not easily determine how likely a threat has to be to be non-negligible. Despite this, many of us are able to make coherent judgments about these matters. The preceding formulation seems to be sufficiently precise to have required minimizing the use of trichloroethylene in the 1990s. The threat that trichloroethylene could be a human carcinogen was non-negligible, and the cost of avoiding or reducing this threat was not unreasonable. Therefore, it seems sensible to conclude that the regulators should have acted earlier than they did.

RISK AND INFORMED CONSENT

The risk-benefit principle and the precautionary principle draw the line between acceptable and unacceptable risks by considering the risk's potential consequences as well as the information we have (or do not have) about the probability of those consequences. A fundamentally different approach is to argue that the properties of the risk itself are morally irrelevant. On this alternative view, what matters is whether those

CASE 10-2

Spaghetti Code in Toyota's Software?

On August 28, 2009, Mark Saylor was driving three family members in a Lexus ES 350 on a highway in San Diego. Mr. Saylor, forty-five, was an off-duty highway patrol officer and vehicle inspector. Even though Mr. Saylor did not press the gas pedal, the car suddenly accelerated to over 100 mph. The car did not stop when he pumped the breaks. One of the passengers in the car called 911 to ask for help, but it was too late. The Lexus eventually hit another vehicle, rolled over, and burst into flames. No one survived the crash.

Between 2000 and 2009, the National Highway Traffic Safety Administration (NHTSA) received more than 6,200 complaints concerning unintended accelerations in Toyota Corollas, Camrys, and various Lexus models. (Lexus is a subsidiary of Toyota and shares some technology with the parent company.) It is believed that unintended acceleration was the cause of at least eighty-nine deaths and fifty-seven injuries. Toyota investigated the issue and concluded that the root cause was a heavy floor mat that sometimes trapped the gas pedal. Eight million vehicles were recalled in 2009. Less than a year later, Toyota also admitted that the gas pedal was made of an unusually sticky material, which triggered a second recall. Despite all this, many victims refused to believe that the floor mat or gas pedal was the true root cause. Some affected drivers felt there must be something wrong with the electronic system that controls the throttle. In the affected Toyota models, just as in other modern cars, the throttle is no longer controlled by a mechanical wire but rather by an electronic drive-by-wire system.

In March 2010, the US Department of Transportation asked a team of thirty NASA engineers to study the electronic system in Toyota's Engine Control Unit (ECU). After ten months of investigation, NASA reported that they "found no evidence that a malfunction in electronics caused large unintended accelerations." The Secretary of Transportation told the press that "We enlisted the best and brightest engineers to study Toyota's

electronics systems, and the verdict is in. There is no electronic-based cause for unintended high-speed acceleration in Toyotas." Despite this unambiguous conclusion, some victims and experts refused to believe that the ECU was working properly. They began to question whether Toyota had given NASA correct and complete information.

A breakthrough came in 2013 when a court in Oklahoma ruled that "it was more likely than not" that Toyota had not taken "reasonable care" when designing the ECU. In the lawsuit, Ms. Jean Bookout, seventy-six, claimed that her 2005 Toyota Camry had accelerated unexpectedly and unintentionally, which made her car crash into an embankment. Ms. Bookout survived, but her passenger and friend Barbara Schwartz, seventy, died. The floor mat on Ms. Bookout's Camry was of a different type, so this accident could not be blamed on that. Toyota's counsel speculated that Ms. Bookout might mistakenly have pressed the gas pedal instead of the brake pedal.

The plaintiff's attorney called two expert witnesses who described the software in Toyota's ECU as a "spaghetti code" with more than 11,000 global variables. They found sixty-seven functions in the code with a cyclometric complexity over fifty (this is a measure of software complexity). This means the code was "untestable." Moreover, the cyclometric complexity of the throttle angle function, which was crucial for controlling the throttle, was one hundred. This means that the code was "unmaintainable." Any modification to one part of the function was likely to trigger unintended and unforeseeable effects elsewhere in the program, with unknown consequences.

Despite all this, the experts did not find any single bug they *knew for sure* would lead to unintended acceleration, although they found many possible candidates. They identified over eighty thousand violations of the software standard set by the Motor Industry Software Reliability Association, as well as many violations of Toyota's internal software standards. The experts also discovered that about 94 percent of the memory stack in the ECU was used during normal operation. The experts decided to figure out if there was a risk the ECU could run out of memory, and what would happen to the throttle control if it did. Their conclusion was not flattering for Toyota. It was highly likely that the ECU would run out of memory; and under such conditions, the throttle would not be reset to zero but rather remain constant or increase a little bit every time the error occurred. To put it briefly, the ECU had not been designed to *fail safely*. (In cars with a wire attached to the gas pedal, a snapped wire will automatically rest the throttle to zero. This is a paradigmatic example of how to design things in ways that take into account what will happen when the technology fails.)

The jury in Oklahoma awarded $3 million in compensatory damages to Jean Bookout and the family of Ms. Schwartz; but before any punitive damages could be determined, the case was settled out of court. Toyota did not admit that the software in the ECU was to blame for the accident. However, about a year later, Toyota agreed to pay a $1.2 billion fine to the federal government for having withheld information and misled government officials about the many reports of unintended acceleration.

There are several ethically relevant points to be made about the Toyota case, including points about risk. For instance, at the beginning of 2010, the regulators did not know for sure if there was a problem with the ECU. However, the NHTSA had received thousands of complaints, and the Department of Transportation took the issue seriously enough to ask NASA to investigate the electronic systems in the affected models. This suggests that it might have been reasonable to suspect already at this point that there was a problem with the ECU. If so, it might have been relevant to ask if the precautionary principle should have been applied.

According to the Wingspread Statement on the precautionary principle, "precautionary measures should be taken even if some cause and effect relationships are not fully established scientifically," provided that "an activity raises threats of harm to human health." According to this formulation, there was no need to *prove* that the software in the ECU was responsible for unintended acceleration. What matters is whether it was *reasonable to believe* that the ECU was responsible for numerous deaths and injuries. The fact that the NASA engineers found "no evidence" of any safety-critical bugs during their investigation in 2010 does not entail that there were none, and the many reports of mysterious accidents suggested the ECU could be to blame. So even if it was far from certain in 2010 that the software was responsible for the accidents, it might nevertheless have been reasonable to believe so. It is better to be safe than sorry.

Discussion question: Try to apply the engineering definition of risk to the Toyota case. What missing information would you need for successfully performing a quantitative risk-benefit analysis? How likely is it that it would be possible to obtain this information? Is the mere fact that it seems *easier* to apply the precautionary principle to the Toyota case a reason for doing so?

---

exposed to the risk have given their *informed consent* to being exposed to the risk. The core idea in this libertarian approach can be stated as follows:

> It is morally permissible to impose a risk on other people just in case they have given their informed consent.

The principle of informed consent is widely accepted across all subfields of applied ethics, most notably in medical ethics. A doctor must obtain the patient's informed consent before performing any medical intervention. The cruel medical experiments carried out by Nazi doctors on prisoners in concentration camps during World War II illustrate the consequences of medical interventions performed without informed consent.

According to the principle of informed consent, it is not the magnitude of the risk that determines its acceptability. If you make an informed decision to voluntarily play Russian roulette with me, and you are fully aware that the probability that you will die is 1 in 6, then it is strictly speaking not unethical to expose you to this risk. You accepted the risk and were fully informed of the possible consequences. However, doctors would never tolerate medical products that exposed healthy patients to such risks, no matter what the patients themselves thought about the risk. What is lacking is a clause that prevents us from accepting a clearly irrational risk. We might formulate the principle better as so:

> It is morally permissible to impose a risk on other people just in case they have given their informed consent and the decision to accept the risk is not clearly irrational.

On either version of the principle, the notion of consent will not always apply neatly to engineering decisions. Astronauts who participate in risky space missions typically make well-informed decisions. They receive plenty of information, fully understand the information they receive, and can freely decide whether to accept missions. But this does not apply to everyone exposed to technological risks. If a nuclear reactor is built where you live, it is unlikely that you can fully understand the information you receive; and even if unwilling to consent to the risk, there is little you can do about it. You should not have to sell your house and move hundreds of miles away just because you do not consent to the risk, yet it would be virtually impossible to build new nuclear power plants, highways, oil rigs, and airports if builders had to obtain informed consent from everyone at risk.

An additional problem is that often it is impossible to ask virtually everyone at risk for informed consent. Imagine, for instance, that you have developed a new type of asphalt that you wish to test on Interstate 45 between Houston and Dallas. Because people who use the highway will be participating in an experiment with an unknown outcome (the new asphalt may turn out to be very slippery), you need to obtain informed consent from every driver. type of do you do this? Do you stop each and every car entering the highway to ask them to sign a waiver? If so, what if they decline participation in your asphalt experiment? Will you build a parallel highway all the way to Dallas? That would be very costly and impractical. While desirable to obtain informed consent whenever feasible and not unreasonably costly, there will always be cases where it is infeasible. In those cases, engineers may not be required to respect the principle of informed consent. An alternative approach could be to ask a representative sample of individuals what they would be willing to consent to, or determine what a fully informed and rational individual (whose values may differ in significant ways from your own) would be willing to consent to.

RISK AVERSION

The engineering approach to risk discussed at the beginning of this chapter is risk neutral. To put it briefly, the risk-benefit principle entails that very large risks can always be balanced against other sufficiently large benefits. This does not always tally well with our intuitions about risk aversion.

The best way to explain why the engineering approach to risk-benefit analysis fails to account for intuitions about risk aversion is to consider a famous example involving

CASE 10-3

Germany and the Fukushima Accident in Japan

On March 11, 2011, a tsunami caused by a magnitude 9.0 earthquake led to meltdowns in three of the six nuclear reactors in Fukushima, Japan (see Figure 10.3). As one might have expected, the engineers who designed the reactors had taken the possibility of earthquakes into account and designed the Fukushima reactors to stop automatically. Initially, the automatic stop system worked as intended. However, the 128 ft. (39 m) tsunami wave that followed the earthquake was higher than the reactor facility could withstand. The emergency generators needed for cooling the reactors were flooded with sea water and stopped, which led to meltdowns in the reactors and the release of radioactive material into the air and sea. The direct cause of the disaster was the lack of electricity for

cooling the reactors during the automatic shutdown process.

The events in Fukushima triggered an extensive public debate over nuclear power in Germany. In 2011, nuclear power accounted for 22 percent of its national electricity supply. Pressed by the media and influential public intellectuals, German chancellor Angela Merkel announced in May 2011 that half of the nuclear reactors in Germany would be shut down immediately and the remaining ones phased out by 2022. The decision was taken before the official Japanese report on the accident had been published. Because German reactors are not susceptible to earthquakes or tsunamis, this radical response to the Japanese accident surprised many observers.

Figure 10.3
On Friday, March 11, 2011, a 9.0 magnitude earthquake off the coast of Japan killed at least 15,894 people and destroyed 127,290 buildings. An additional one million buildings were damaged. There have been no reported cases of radiation sickness linked to the meltdowns in the Fukushima Daiichi reactors.
Source: iStock by Getty Images.

(Continued)

Was it right or wrong to phase out nuclear power in Germany in response to the accident in Japan? This is, of course, difficult to say. Reasonable people can disagree on this, even if they have access to the same factual information. However, it seems clear that it would have been a mistake to use the precautionary principle for warranting a ban on nuclear power. The objective risk in Germany remained the same as it was before the accident in Fukushima. It was only the subjective risk that increased. If it was ethically permissible to use nuclear power in Germany before the Fukushima accident, it was permissible to do so after the accident too. From an ethical perspective, nothing significant changed because of the accident—except for the public's risk perception. German nuclear reactors did not become less safe from one day

to another because of the meltdowns in Japan. If so, the German ban on nuclear power two months after the accident in Japan does not satisfy the clause in the precautionary principle requiring precautionary measures to be "reasonable." However, the political decision to phase out nuclear power in Germany was perhaps a reasonable *democratic* decision considering the change in people's risk perception.

Discussion question: What is the role of experts in democratic societies seeking to regulate new and existing technological risks (such as nuclear power)? Should those who understand the technology, that is, engineers and other technical experts, decide whether a risk is acceptable; or is the role of the expert to merely provide information to politicians and other democratically elected representatives who are ultimately responsible for regulatory decisions?

a single decision maker asked to choose among a set of lotteries. The example is due to the French economist and Nobel Prize winner Maurice Allais, who asks us to consider the four lotteries in Table 10.2 in which exactly one winning ticket will be drawn.

In a choice between Lottery 1 (L1) and Lottery 2 (L2), it seems reasonable to prefer L1 because if you take L2, there is a risk you end up with nothing. If you choose L1, it is certain that you win $1 million. So, intuitively, a risk-averse decision maker would choose L1. However, in a choice between L3 and L4, it seems reasonable to trade a ten in one hundred chance of getting $5 million rather than $1 million against a one in one hundred risk of getting nothing instead of $1 million. So it would not be unreasonable to prefer L4.

What has this got to do with risk aversion? No matter how much or how little you care about money, the risk-benefit principle recommends you maximize expected value. This entails that you *must* prefer L1 to L2 if and only if you prefer L3 to L4. There is no way of assigning value to money (which is consistent with the expected value principle) that permits you to prefer L1 to L2 and L4 to L3. To see why, we shall calculate the difference in expected value between the lotteries. The probability that ticket 1

Table 10.2 No matter how much or little you care about money, the principle of maximizing expected value entails that you must prefer Lottery 1 to Lottery 2 if and only if you prefer Lottery 3 to Lottery 4.

	Ticket 1	Ticket 2–11	Ticket 12–100
Lottery 1	$1,000,000	$1,000,000	$1,000,000
Lottery 2	$0	$5,000,000	$1,000,000
Lottery 3	$1,000,000	$1,000,000	$0
Lottery 4	$0	$5,000,000	$0

will be drawn is 0.01, the probability that one of the tickets numbered 2–11 will be drawn is 0.1, and the probability that one of the tickets numbered 12–100 will be drawn is 0.89. Let exp(L1) denote the expected value of L1 and $v(1M)$ the agent's personal value or utility of \$1 million, and so on. We now have the following:

$$exp(L1) - exp(L2) = v(1M) - [0.01v(0M) + 0.1v(5M) + 0.89v(1M)] = 0.11v(1M) - [0.01v(0) + 0.1v(5M)]$$

$$exp(L3) - exp(L4) = [0.11v(1M) + 0.89v(0)] - [0.9v(0M) + 0.1v(5M)] = 0.11v(1M) - [0.01v(0) + 0.1v(5M)]$$

These equations show that the difference in expected value between L1 and L2 is *precisely the same* as the difference between L3 and L4. So no matter how much or how little value you place on money, you cannot simultaneously prefer L1 to L2 *and* prefer L4 to L3 without violating the expected value principle.

The problem with all this is that the engineering definition of risk seems to give the wrong recommendation. A reasonable attitude to risk is to prefer L1 and L2 *and* L4 to L3, but the engineering definition does not permit us to do so. We cannot fix this problem by simply adjusting the values we assign to the outcomes. The problem is deeper than that. The fact that one of the options is *risk free* is something that seems to be of a very fundamental importance to our normative intuitions, in a way that cannot be captured by the expected value principle.

REVIEW QUESTIONS
1. Explain the distinction between subjective and objective risk.
2. Explain the engineering definition of risk.
3. Why was it difficult to apply the engineering definition of risk to the trichloroethylene case?
4. How does a risk-benefit analysis differ from a cost-benefit analysis?
5. Explain, by using your own words, the main ideas behind the precautionary principle.
6. Must defenders of the risk-benefit principle reject the precautionary principle and vice versa?
7. What were the root causes of the Fukushima accident in Japan?
8. What role, if any, should the principle of informed consent play in risk management?

REFERENCES AND FURTHER READINGS

Ashford, N., K. Barrett, A. Bernstein, R. Costanza, P. Costner, C. Cranor, P deFur, et al. 1998. "Wingspread Statement on the Precautionary Principle." http://www.psrast.org/precaut .htm. Accessed November 2, 2015.

Cusumano, M. A. 2011. "Reflections on the Toyota Debacle." *Communications of the ACM* 54, no. 1: 33–35.

General Assembly of the United Nations. 1992. "Report of the United Nations Conference on Environment and Development." *A/CONF* 151, no. 26. New York: United Nations.

Kirsch, Michael T., V. A. Regenie, M. L. Aguilar, O. Gonzalez, M. Bay, M. L. Davis, C. H. Null, R. C. Scully, and R. A. Kichak. January 2011. "Technical Support to the National Highway Traffic Safety Administration (NHTSA) on the Reported Toyota Motor Corporation (TMC)

Unintended Acceleration (UA) Investigation." *NASA Engineering and Safety Center Technical Assessment Report.* Washington, DC: NHTSA.

National Toxicology Program. 2011. "Trichloroethylene." In *Report on Carcinogens,* 12th ed. Washington, DC: United States Department of Health and Human Services, 420–423.

Peterson, M. 2007. "Should the Precautionary Principle Guide Our Actions or Our Beliefs?" *Journal of Medical Ethics* 33, no. 1: 5–10.

Peterson, M. 2017. *The Ethics of Technology: A Geometric Analysis of Five Moral Principles.* New York: Oxford University Press.

Steel, D. 2014. *Philosophy and the Precautionary Principle.* Cambridge, UK: Cambridge University Press.

Svensson, S., and S. O. Hansson. 2007. "Protecting People in Research: A Comparison between Biomedical and Traffic Research." *Science and Engineering Ethics* 13, no. 1: 99–115.

Taebi, B., and S. Roeser, eds. 2015. *The Ethics of Nuclear Energy.* Cambridge, UK: Cambridge University Press.

NOTES

1. The American Cancer Society writes on its website that "the [radiofrequency] waves given off by cell phones don't have enough energy to damage DNA directly or to heat body tissues"; but the organization does not make any definitive statement about what we know about this potential cancer risk. https://www.cancer.org/cancer/cancer-causes/radiation-exposure/cellular-phones.html

2. National Toxicology Program, "Trichloroethylene," *Report on Carcinogens*, 12th ed. (Washington, DC: United States Department of Health and Human Services, 2011), 420.

3. General Assembly of the United Nations, "Report of the United Nations Conference on Environment and Development," *A/CONF* 151, no. 26 (New York: United Nations, 1992), 12.

4. Ashford et al., "Wingspread Statement on the Precautionary Principle," PSRAST website, http://www.psrast.org/precaut.htm (1998). Accessed November 2, 2015.

Privacy: What Is It and Why Should It Be Protected?

Technological devices sometimes collect and store sensitive private information about shopping habits, personal friendships, political views, and lifestyle choices. Web browsers, cell phones, and CCTV systems are obvious examples. Many people worry about this. Are we building a society similar to that described in George Orwell's dystopic novel *1984*? If so, do engineers have an obligation to develop technologies that protect our privacy?

Interestingly, the worry that new technologies pose threats to people's privacy is by no means a modern phenomenon. In 1890, Louis Brandeis and Samuel Warren published an article in *Harvard Law Review* entitled "The Right to Privacy." They claimed that

> Instantaneous photographs and newspaper enterprise have invaded the sacred precincts of private and domestic life; and numerous mechanical devices threaten to make good the prediction that "what is whispered in the closet shall be proclaimed from the house-tops."[1]

Brandeis and Warren argued that citizens have a (legal) right to privacy, but they admitted that it is difficult to see how this right could be derived from the Lockean conception of natural rights, according to which we own ourselves and the fruits of our labor (see chapter 6). Suppose, for instance, that I take a photo of you, without violating any of your other rights, and sell it to a newspaper. Because I am the legitimate owner of the photo, the mere fact that you own yourself does *not* entail that you are entitled to stop me from selling the photo. Your ownership right of yourself is not violated if I use a technological device that collects information about you, given that I do so without infringing on any of your other rights. If you voluntarily decide to go for a walk downtown in an ugly green and purple suit, I am permitted to take a photo of you wearing your silly outfit in public.

Considering these examples, the intellectual challenge we face in this chapter is to explain what privacy is and why and when it should be protected. Before we go into the philosophical details, it is helpful to consider a more recent case in which concerns about privacy and anonymity clearly need to be balanced against other moral concerns—in particular, the right to free speech.

CASE 11-1

Google and the Right to Be Forgotten

In 2010, Mr. Mario Costeja González, a Spanish citizen, searched for his own name in Google's search engine. Google directed Mr. González to a web page in which it was stated that his home had been foreclosed and sold at an auction in 1998. This information was correct. Mr. González's home had indeed been sold by his mortgage lender because Mr. González had been unable to make the required interest payments. However, Mr. González felt that this information was no longer relevant because he had recently paid off his debt. In Mr. González's opinion, the information about his foreclosed home was a violation of his privacy.

After four years of legal processes, the Court of Justice of the European Union ruled that "Individuals have the right—under certain conditions—to ask search engines to remove links with personal information about them."[2] The court ordered Google (see Figure 11.1) to remove all websites that contain information about the man's foreclosed house from its index.

Within a couple of months Google received thousands of similar requests from citizens all over Europe to remove other links. The company quickly implemented a standardized procedure for dealing with the requests and removed at least 1,390,838 pages from its index. On Google's European website (Google.de, Google.it, Google.es, etc.) the company explains at the bottom of page that "Some results may have been removed under data protection law in Europe." Yahoo, Bing, and other search engines have implemented similar procedures and warning messages.

Figure 11.1
The Court of Justice of the European Union has ruled that individuals have the right to ask search engines to remove links with personal information.
Source: iStock by Getty Images.

Is it morally right to filter search results that contain sensitive personal information? It is easy to see that doing so may benefit the individual whose privacy is at stake. At the same time, it is not so easy to see what the moral basis for this request for privacy could be. Freedom of speech is a basic human right according to the United Nation's Universal Declaration of Human Rights. So if we wish to grant citizens a right to privacy, we have to somehow make sure that it does not conflict with other fundamental rights.

Discussion question: What determines what pieces of information one could sometimes legitimately be asked to keep private? Is it the mere fact that you do not *want* others to know certain things about you sufficient for invoking a justified claim to privacy, or must some additional criteria be met? Is there an objective standard for what counts as "sensitive" information?

PRIVACY AS A MORAL RIGHT

Do we have a right to privacy? As explained by Brandeis and Warren in their time-less piece, we cannot base the right to privacy on the Lockean conception of natural rights because privacy is not a property right. To bring home this point, Brandeis and Warren discuss a court case from 1888 in which "a photographer who had taken a lady's photograph under the ordinary circumstances was restrained from exhibiting it, and also from selling copies of it, on the ground that it was a breach of an implied term in the contract."[3]

Let us suppose for the sake of the argument that there was, in this particular case, an implied term in the contract that prevented the photographer from selling the sensitive photo. Could such a notion of implicit agreement be the basis for an alleged right to privacy? Brandeis and Warren argue that in many other cases, the notion of implied agreement would be insufficient for generating reasonable conclusions about privacy. Suppose, for instance, that a stranger positions himself on the street in front of your house. He takes a photo from the street of you and your partner in the bedroom under what we may euphemistically describe as "the ordinary circumstances." In this case, there would be no implied agreement between you and the photographer, but it nevertheless seems reasonable to maintain that the stranger violated your privacy.

Another example could be a case in which a stranger accidently receives an intimate email intended for someone else. Although the accidental recipient of the email made no prior explicit or implicit commitment to keep the content of the email confidential, it seems reasonable to claim that not doing so would be a violation of the sender's privacy.

Brandeis and Warren's solution is to introduce the right to privacy as a new, free-standing right. From a legal point of view, this is a simple fix. However, from an ethical point of view, this maneuver seems somewhat arbitrary. If we cannot derive all rights from a single criterion (in natural rights theory: Locke's theory of ownership), how can we then be sure we actually have a moral right to privacy? And why stop with the right to privacy; why not introduce other rights too? We could, for instance, stipulate that everyone has a right to a happy marriage, or a right to five fantastic vacations per year. To put it briefly, the worry is that if we base new rights on "thin air," then the notion of rights runs the risk of being deprived of its moral force.

In the Universal Declaration of Human Rights ratified by the United Nations in 1948, human rights are related to the concept of *human dignity*. The UN claims that "recognition of the inherent dignity and of the equal and inalienable rights of all members of the human family is the foundation of freedom, justice and peace in the world."[4] On this view, it could be argued that respect for human dignity requires a right to privacy. If we do not respect people's right to be forgotten on the Internet, we sometimes do not respect their human dignity. A weakness of this analysis is, of course, that the concept of human dignity is somewhat opaque. Exactly what is human dignity and how can we tell if our human dignity requires a right to privacy?

An alternative approach is to conceptualize the right to privacy as a socially constructed right. On this view, the moral justification of the right to privacy ultimately relies on the consequences of respecting people's privacy. We have a right to privacy just in case a society in which people's privacy is respected is a better one than a society in which we don't respect privacy. The key term "better society" can be spelled out in various ways. Utilitarians stress the importance of maximizing well-being. Virtue ethicists focus on whether the right to privacy helps to promote virtues such as prudence, temperance, justice, and courage.

If we conceive of privacy as a socially constructed right, it is relatively easy to understand why the European Court of Justice concluded that the right to be forgotten can sometimes outweigh what many Americans would consider to be an infringement of the right to free speech. The European view is that all things considered, a society in which search engines have to remove sensitive links is a better society, even if this sometimes infringes on other rights.

A possible worry with the social constructivist approach is that it makes the right to privacy contingent on empirical facts. Consider, for instance, the question whether cheap, private, camera-equipped drones should be allowed to take photos of people's backyards. In areas where the crime rate is high, the overall effects of allowing privacy-invasive drones might be positive; but in rich neighborhoods with low crime rates, the empirical facts might be very different. It seems odd to maintain that one's right to privacy in a given neighborhood should depend on the effect such a right would have on the crime rate in that particular neighborhood.

PRIVACY AS A MORAL VALUE

Rather than claiming that people have a *right* to privacy, it could be argued that privacy is an important *moral value*—just like justice, freedom, and happiness. The difference between rights and values is that the former tend to be binary in ways that values are not. You either have a right or you don't, and every violation of your moral rights is morally wrong. However, moral values vary in strength and typically have to be balanced against other conflicting values. Therefore, if privacy is conceptualized as a value rather than a right, it will not always be morally wrong to violate people's privacy. There may exist situations in which it is permissible to sacrifice privacy to bring about larger quantities of other values such as justice, freedom, or happiness. Imagine, for instance, that the NSA violates our privacy by reading sensitive, private emails. If this enables the agency to gather information that prevents terrorist attacks, such privacy intrusions could perhaps be acceptable. If we gain sufficiently large quantities of happiness and freedom by preventing terrorist attacks, it may not be wrong to sacrifice some privacy.

A potential weakness of this line of reasoning is that it is not entirely clear *why* privacy would be a moral value. Why does the private sphere deserve special protection? To explain this, it is helpful to compare two hypothetical examples. In the first, a shop owner uses CCTV cameras for monitoring your behavior as you buy a bottle of red wine late at night. You are not drunk and you do nothing illegal; and because you have nothing to hide, you do not worry about the fact that your visit is recorded on video. Compare this to an example in which a potential terrorist is being monitored by the FBI for months. All the suspect's web searches, credit card transactions, and physical movements are being carefully recorded day out and day in. However, after a year, it turns out that the potential terrorist was completely innocent. He is a lawful citizen who has never done anything illegal or morally questionable.

If we believe privacy is a moral value on a par with justice, freedom, and happiness, we have to claim that it would somehow have been better if the shop owner and the FBI had not used any privacy-intrusive technologies. In both examples, the subjects did nothing wrong, so the potential gain of monitoring their behavior never materialized. Therefore, it seems appropriate to ask: What is it that is lost if we sacrifice some privacy in a (failed) attempt to promote other values?

It is sometimes argued that lawful, innocent citizens have nothing to fear from privacy-invasive technologies. On that view, if you do nothing wrong, you have no reason to worry if Big Brother is watching you. However, it is worth keeping in mind that the value of privacy is intimately linked to autonomy. If we do not respect people's privacy, they are less likely to be fully autonomous. Nearly everyone agrees that autonomy is an important moral value. For instance, according to Kant, "a will whose maxims necessarily coincide with the laws of autonomy is a holy will, good absolutely."[5] To see how privacy-invading technologies could make fully innocent and lawful citizens less autonomous, we can imagine that the customer in the wine shop was a well-known politician who would actually have preferred to buy six bottles of wine instead of one (with the intent to drink one bottle per month for the coming six months); the reason he bought just one bottle was because he knew his visit to the store was being recorded on video, and the shop owner could leak it to the press. For a politician, it does not look good to buy too much alcohol on a single day.

The point of this example is that fully innocent and lawful citizens may *sometimes adjust their behavior* if privacy-invasive technologies are used. This makes them less autonomous.

The same point applies to the terrorist example. A person falsely accused of planning a terrorist attack is likely to adjust his behavior if he suspects the FBI is watching him. Although he might have had a perfectly legitimate reason for buying dynamite, he is likely to refrain from doing so. This, again, shows why privacy-invasive technologies make us less autonomous. The agent's behavior will not reflect his true desires and preferences, even in cases in which those desires and preferences are perfectly legitimate. The agent does not freely choose whatever he would have chosen had the technology not been used.

WHAT CAN ENGINEERS DO TO PROTECT PRIVACY?

Sometimes the best way to address a moral issue is to make sure it never arises. Engineers can often protect people's privacy when they *design* technological systems. The following are some examples of how engineers can address privacy concerns at the design stage. What these and other examples show is that the solution to some, but not all, privacy concerns is to build technological systems in ways that better protect our privacy.

Full-Body Scanners

In many airports around the world, full-body scanning technology is used for detecting objects concealed underneath people's clothing. When the technology was first introduced in 2008, the scanners displayed a detailed image of the surface of the person's skin. However, in 2013, the US Congress prohibited the use of such privacy-invasive image technologies. In response to this legal ban, engineers redesigned the scanners. They currently use a system called Automatic Target Recognition (ATR), which displays a generic body outline instead of the person's actual skin; and some parts of the body are masked by the software (see Figure 11.2).

VPN

The abbreviation "VPN" stands for Virtual Private Network. VPN networks allow users to communicate with each other through a virtual private "tunnel" on the Internet. If you, for instance, visit China and wish to access your Gmail account (which is blocked by the Chinese authorities), you can do so by creating a VPN connection to your server at home. Data sent through a VPN tunnel can still be monitored by the authorities; but because it is encrypted, no unauthorized individuals will be able to decipher the meaning of the transmitted information.

WhatsApp

The WhatsApp messenger software has been designed to collect as little personal information as possible. Jan Koum, founder of WhatsApp, explains:

> Respect for your privacy is coded into our DNA, and we built WhatsApp around the goal of knowing as little about you as possible: You don't have to give us your name

Figure 11.2
This body scanner is designed to minimize privacy intrusions by not revealing more information that necessary to the security personnel.   Source: iStock by Getty Images.

and we don't ask for your email address. We don't know your birthday. We don't know your home address. We don't know where you work. We don't know your likes, what you search for on the Internet or collect your GPS location. None of that data has ever been collected and stored by WhatsApp, and we really have no plans to change that."[6]

Traffic Cameras

Many cities use traffic cameras for recording red light violations and speeding. Ideally, such cameras should just record the license plate of the vehicle but not the identity of the driver or passengers. If you are driving around in town late at night with the "wrong" person, then people monitoring the images have no right to know that. To avoid this concern, some traffic cameras have been designed to take photos from the *rear* of the vehicle instead of the front. This is sufficient for registering the license plate, but it makes it impossible to record the identity of the driver and passengers.

REVIEW QUESTIONS

1. Why is the right to be forgotten acknowledged by the European Union?
2. What are the pros and cons of basing a right to privacy on Locke's claim that we own ourselves and the fruits of our labor?
3. What are the pros and cons of thinking of privacy as a socially constructed right?
4. What are the pros and cons of thinking of privacy as a moral value?
5. What argument could one give for the claim that privacy-invasive technologies may limit the freedom also for agents who have no illegitimate preferences and nothing to hide?
6. Give some examples of what engineers can to do protect people's privacy as they design new technologies.

REFERENCES AND FURTHER READINGS

Bennett, S. C. 2012. "The Right to Be Forgotten: Reconciling EU and US Perspectives." *Berkeley Journal of International Law* 30: 161–195.

European Commission. 2012. "Proposal for a Regulation of the European Parliament and of the Council on the Protection of Individuals with Regard to the Processing of Personal Data and on the Free Movement of Such Data." 2012. Article 17. General Data Protection Regulation, Right to Be Forgotten and to Erasure.

Kant, Immanuel. 2011. *Groundwork of the Metaphysics of Morals: A German-English Edition*. Edited and translated by Mary Gregor and Jens Timmermann. Cambridge, UK: Cambridge University Press. First published 1785.

Koum, J. March 17, 2014. "Setting the Record Straight." WhatsApp.com (blogpost). https://blog.whatsapp.com/529/Setting-the-record-straight. Accessed May 25, 2017.

Warren, S. D., and L. D. Brandeis. 1890. "The Right to Privacy." *Harvard Law Review* 4: 193–220.

NOTES

1. Samuel D. Warren and Louis D. Brandeis, "The Right to Privacy," *Harvard Law Review* (1890): 195.
2. European Commission, "Proposal for a Regulation of the European Parliament and of the Council on the Protection of Individuals with Regard to the Processing of Personal Data and on the Free Movement of Such Data" (General Data Protection Regulation, 2012), 1.
3. Warren and Brandeis, "The Right to Privacy," 208.
4. UN General Assembly, *Universal Declaration of Human Rights* (New York: UN General Assembly, 1948). The quote is from the first sentence of the preamble.
5. I. Kant, *Foundations of the Metaphysics of Morals*, trans. Lewis White Beck (1785; repr., Indianapolis, IN: Bobbs-Merrill Library of Liberal Arts, 1956), p. 438 in the Prussian Academy edition.
6. J. Koum, "Setting the Record Straight," WhatsApp.com (blogpost), March 17, 2014, https://blog.whatsapp.com/529/Setting-the-record-May 25, 2017.

CHAPTER 12

The Problem of Many Hands: Who Is Responsible and Should Anyone Be Blamed?

Engineers often work in teams. If something goes wrong (a bridge collapses, a river is polluted, some critical piece of software crashes), it is not uncommon that no single engineer can be held responsible or blamed for the negative outcome. Because engineering projects are performed jointly by several technical experts, the responsibility for negative outcomes is often shared by several team members. This brings us to the main question of this chapter: How should moral responsibility be distributed across a set of individuals when several agents jointly perform actions that lead to a morally unacceptable outcome? In our discussion, we will pay special attention to a phenomenon known as *the problem of many hands*, which occurs whenever there is a "gap" in the distribution of responsibility within a group of agents. The infamous Hyatt Regency walkway collapse in Kansas City is a well-known example. (See case 12-1.) Although no single individual was responsible for the disaster, we nevertheless have reason to believe that *someone* did something he or she should be held responsible for, although it is very difficult (and perhaps impossible) to specify each agent's degree of responsibility.

WHAT IS RESPONSIBILITY?

The concept of responsibility is discussed in some detail by Aristotle in the *Nicomachean Ethics*. He argues that we are responsible for our individual actions just in case two conditions are fulfilled. First, we must be able to *voluntarily decide* whether to perform the action. Second, we must be *aware* of what we are doing and have accurate information about the relevant circumstances. A consequence of the first condition (voluntariness) is that a slave who is ordered to perform some cruel action in a prison camp is not responsible for that action. A consequence of the second condition (awareness) is that a drunk person is not responsible for actions carried out under the influence of alcohol.

To avoid the unfortunate implication that we cannot be held responsible for any wrongdoing when drunk, Aristotle makes room for the plausible idea that we are

CASE 12-1

The Hyatt Regency Walkway Collapse

Built in steel, concrete, and glass, the forty-story Hyatt Regency Kansas City Hotel opened after two years of construction in 1980. The lobby incorporated a large open atrium with multiple walkways connecting the south and north wings on the second, third, and fourth floors. The walkways on the second and fourth floors were located directly above each other.

On July 17, 1981, about 1,600 people attended a "tea dance" in the atrium. Approximately 40 attendees were standing on the walkway on the fourth level and 20 on the second level walkway. At 7:05 p.m., both walkways collapsed and fell onto the crowd in the lobby, killing 114 people and injuring 216. This is the deadliest *accidental* structural disaster in US history. (The collapse of the World Trade Center in New York on September 11, 2001, was not accidental.)

The second floor walkway was suspended in 1.25 in. steel rods from the fourth floor walkway, which were suspended in rods of the same dimension from the ceiling. The direct cause of the collapse was a change of the design carried out by the contractor that effectively doubled the load on the nut holding the fourth floor walkway (see Figure 12.1). The walkways were designed by Jack D. Gillum and Associates and manufactured by Havens Steel Company. During the construction process, Havens Steel Company realized that it would be easier to manufacture the walkways if they replaced the single continuous steel rods that were supposed to hold up both walkways with two separate rods connected as shown in Figure 12.1. Everyone with basic knowledge of statics can verify that this doubled the load on the nut holding the fourth floor walkway.

There is no consensus on how, or if, the design modification proposed by Havens Steel Company was approved by Jack D. Gillum and Associates. In sworn testimony, representatives of Havens claimed that they had called Gillum to ask if the change was acceptable. According to Havens's version, the engineer they spoke to at Gillum accepted the change request without even looking at the

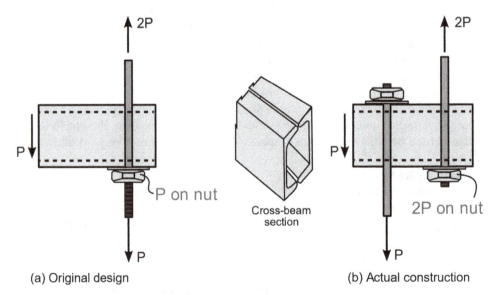

(a) Original design		(b) Actual construction

Figure 12.1
By not using a continuous rod as originally designed, the load on the nut doubled from P to 2P.
Source: Wikimedia.

drawings or performing any calculations. Havens also claimed, and this is not disputed, that they received new drawings from Gillum with the structural engineer of record's review stamp. However, representatives of Gillum told the investigators that they never received any phone call from Havens and that the documents sent to Havens were preliminary drawings. It remains unclear who put the review stamp on those documents. Gillum also stressed that Havens should have contacted Gillum to ask for further clarifications.

Who was responsible for the collapse of the walkways? Based on what we know today, it seems likely that *several* engineers should be held responsible for the disaster. Mr. Gillum at Gillum and Associates had about ten associates working for him, each of whom supervised six to seven projects. Mr. Gillum could not personally oversee sixty to seventy projects, but he was responsible for making sure the company had proper procedures in place for documenting and processing change requests. There is reason to believe that Gillum's administrative processes were inadequate.

The associate responsible for the Hyatt Regency project, Mr. Duncan, was responsible for ensuring that the documents sent to Havens were correct. If (and only if) he approved the change via telephone without looking at the documents, he was of course responsible for that; but even if he did not, it could be argued that he was partly responsible for not having established better and more formal channels of communication with the contractor.

Finally, the supervisor at Havens who made the change request (if such a request was ever made) should not have accepted a verbal approval of the change order. The claim that the modification was verbally approved "with the understanding that a written request for the change would be submitted for formal approval" was largely irrelevant because no such written request was ever submitted.[1] It should also be mentioned that Havens Steel Company subcontracted most of the work on the walkways to a subcontractor, and that company was never informed that the construction was a preliminary design that had not been formally approved by Gillum.

Discussion question: A possible view to take about the Hyatt Regency walkway collapse is that everyone who could have done something to *prevent* the disaster is morally responsible for not having done so. What is, in your opinion, the best objection to this causal account of moral responsibility?

---

typically responsible for *placing ourselves* in the situation in which we are not aware of what we are doing. If you do something wrong when drunk, the wrongdoing occurred earlier, that is, at the point when you voluntarily decided to have too much to drink.

Legal philosopher H. L. A. Hart has pointed out that the term "responsibility" often serves as an umbrella term for a wide range of slightly different notions of responsibility. Hart identifies the following four core notions as particularly interesting:

1. Role-responsibility
2. Causal responsibility
3. Capacity-responsibility
4. Liability-responsibility
   a. Legal liability-responsibility
   b. Moral liability-responsibility

Role-responsibility is something an agent has in virtue of having a certain role. Edward J. Smith was the captain of the *Titanic* when it collided with an iceberg and sank in the North Atlantic in 1912 (see the "The *Titanic* Disaster," Case 12-2). Captain Smith thus had the role-responsibility for the disaster, just like modern-day CEOs have the role-responsibility for nearly everything that happens in their organizations. However, Smith did not become aware of the collision with the iceberg until after it had happened.

Several witnesses testified that Smith was not on the bridge when the ship collided with the iceberg; he was called to the bridge by first officer Murdoch shortly after the collision. What we learn from this is that Aristotle's conditions for responsibility, voluntariness and awareness, apply only to *some* of the four notions of responsibility identified by Hart. Captain Smith had the *role-responsibility* for the disaster even though Aristotle's awareness condition was not satisfied, just as parents have the role-responsibility for their children even if they are not aware of what their children are doing.

To illustrate the four notions of responsibility, Hart constructs an instructive tale about a drunk captain. The numbers in brackets (as they are not included by Hart in the original text) refer to the four notions just listed.

> As captain of the ship, X was responsible [1] for the safety of his passengers and crew. But on his last voyage he got drunk every night and was responsible [4b] for the loss of the ship with all aboard. It was rumored that he was insane, but the doctors considered that he was responsible [3] for his actions. Throughout the voyage he behaved quite irresponsible [4a], and various incidents in his career showed that he was not a responsible person [1, 3]. He always maintained that the exceptional winter storms were responsible [2] for the loss of the ship, but in the legal proceedings brought against him he was found criminally responsible [4a] for the loss of life and property. He is still alive and he is morally responsible [4b] for the deaths of many women and children.[2]

These notions of responsibility are not entirely independent. You cannot, for instance, have the moral or legal liability-responsibility for something unless you also have the relevant mental capacities. Hart also admits that his distinction between causal responsibility and moral liability-responsibility may not reflect the common usage of those terms. In Hart's taxonomy, we should reserve the term moral liability-responsibility for cases in which the agent is alive and is responsible for his action, while causal responsibility refers to cases in which the agent is no longer alive but was responsible for what he or she did.

Without oversimplifying too much, we can say that the common-sense notion of being "morally responsible" for an action refers to situations in which the agent either has some causal responsibility or moral liability-responsibility for the action, and Aristotle's two criteria (voluntariness and awareness) are satisfied.

The common-sense notion of moral responsibility can be divided into *backward-looking* and *forward-looking* responsibility. To establish who is responsible in a backward-looking sense for some negative event, we typically examine the causal history of that event. Backward-looking moral responsibility is thus in many cases the same thing as causal responsibility or moral liability-responsibility. In the *Titanic* example, the lookout and first officer had backward-looking moral responsibility given that Aristotle's awareness condition was fulfilled.

Forward-looking moral responsibility is, as the term suggests, oriented toward the future. You have the forward-looking moral responsibility for some negative event if you are responsible for preventing that negative event from occurring in the future. In the *Titanic* example, the signatories of the International Convention for the Safety of Life at Sea (SOLAS) had the forward-looking responsibility to see to it that similar disasters would not happen again. As explained in the box summarizing the *Titanic* disaster (Case 12-2), SOLAS implemented new safety rules after the *Titanic* sank, which made it mandatory for every ship to carry a sufficient amount of lifeboats for all its passengers and crew.

In the next couple of sections, we will apply the notion of backward-looking responsibility (i.e., causal or moral liability-responsibility) to contexts in which groups or teams of engineers are jointly responsible for some negative outcome. How much of backward-looking moral responsibility should we ascribe to each member of the group?

CASE 12-2

The *Titanic* Disaster

On April 15, 1912, RMS *Titanic* collided with an iceberg and sank in the North Atlantic on her maiden journey from Southampton to New York. The *Titanic* disaster is one of the worst engineering disasters of all times. It is believed that 1,514 of the 2,224 passengers and crew drowned, although the exact figures are somewhat uncertain. At 11:40 p.m. on April 14, lookout Frederick Fleet spotted the iceberg and alerted the bridge. First Officer William Murdoch immediately ordered the engines to be put in reverse and tried to steer away from what he knew would be a major hazard for the ocean liner. The direct cause of the disaster was a series of holes on the starboard side of the hull below the waterline created by the collision with the iceberg.

Naval architect Thomas Andrews had designed the *Titanic* (see Figure 12.2) with sixteen watertight compartments. We now know that if the ship had hit the iceberg with the front of the bow instead of the starboard side, she would not have sunk. The unfortunate last-minute turn allowed five of the watertight compartments to be flooded, which was one too many.

Titanic carried lifeboats for no more than 1,178 passengers and crew. This was less than one-third of her total capacity but in accordance with the safety regulations of the time. More than 80 percent of all children and women in first and second classes survived. However, the death toll for passengers in third class, and for male passengers in all classes, was about four times as high.

Mr. Johan Andersson and his wife Alfrida from Kisa, Sweden, were traveling together with their five children in third class. They were headed for

Figure 12.2
Titanic at the docks of Southampton in 1912. About 1,514 of the 2,224 passengers and crew drowned when ship collided with an iceberg. Who was morally responsible for this disaster?
Source: Wikimedia.

Winnipeg, Canada. Mr. Andersson was one of 387 male casualties in third class (84 percent death rate), while Mrs. Andersson was one of 89 female casualties (54 percent). All five of their children also drowned.

Who was *morally responsible* for the *Titanic* disaster? And who was morally responsible for the death of Mr. and Mrs. Andersson and their five children? Is the person or persons responsible for the sinking of the ship also responsible

(Continued)

for the death of every passenger and crew aboard the ship? Or could different persons or entities be responsible for different "parts" of the disaster?

The answers to these questions depend on what notion of responsibility we have in mind. Are we asking questions about role-responsibility, causal responsibility, capacity-responsibility, or liability-responsibility? Let us, for illustrative purposes, discuss who was causally responsible for the disaster.

If lookout Frederick Fleet had spotted the iceberg earlier, the disaster could have been avoided altogether, but he did not have full causal control of the outcome. The visibility was far from perfect that night, and icebergs were known to be notoriously difficult to spot. If first Officer William Murdoch had kept a straight course into the iceberg, *Titanic* would not have sunk. However, Murdoch did not know that. His intention was to steer away from the iceberg, so he made the wrong decision; but he did not have the right type of causal control because of his limited knowledge of the relevant facts.

Naval architect Thomas Andrews should, perhaps, be held causally responsible to some degree. The sixteen watertight bulkheads were indeed watertight and delayed the sinking by several hours; but if he had specified even more watertight bulkheads, or if they had been watertight all the way up to the deck level instead of just a few meters above the waterline, the ship would not have sunk, or at least not sunk until the nearby ship Carpathia arrived at the scene around 4:00 a.m.

It is also difficult to ignore the role of the maritime authorities that allowed *Titanic* to operate with lifeboats that could carry no more than 1,178 passengers and crew. If there had been enough lifeboats, almost everyone would have survived. One of the most important consequences of the *Titanic* disaster was the creation of the International Convention for the Safety of Life at Sea (SOLAS) passed in 1914, which is still the most important international convention for maritime safety. It has been incorporated in national laws around the world.

Another causal factor that might be relevant to consider is the *cultural norm* that led the crew to prioritize women and children during the evacuation. If this cultural norm had not been in place, it is more likely that Mr. Andersson and the other male passengers would have survived. The same could be said about the cultural norms pertaining to social classes. If Mr. Andersson had been a wealthy business man traveling in first class, instead of a poor immigrant, it is more likely that he would have been given a seat in a lifeboat.

Discussion question: Can we hold *abstract entities* such as cultural norms, social class systems, private corporations, and government agencies responsible (in one or several of the senses outlined previously) for the death of human beings? Or should we rather hold the *individuals* who accept and act in accordance with cultural norms and work for private corporations or government agencies responsible?

THE PROBLEM OF MANY HANDS

The problem of many hands arises in situations in which it is reasonable to conclude that a group of people is jointly responsible for an outcome, although it is difficult or impossible to assign responsibility to each individual member of the group. There is, so to speak, a "gap" between the responsibility assigned to the group as whole and the individual members of the group.

The *Titanic* disaster illustrates some of the key features of the problem of many hands. Many people could have performed actions that would have prevented the disaster: The lookout could have spotted the iceberg earlier; the first officer could have refrained from the last-minute turn; the naval architect could have specified bulkheads that were watertight all the way up to the deck level; the maritime authorities could have insisted on lifeboats for all passenger and crew; and so on. These facts raise the question of whether *everyone* who had some casual influence on the disaster should be held morally responsible for it (in the backward-looking, causal sense). If so, how much moral responsibility should be assigned to each member of the group?

Suppose you believe that everyone's moral responsibility for an outcome is proportionate to the agent's causal influence on the outcome. The greater your causal influence is, the more responsible you are, everything else being equal. In the *Titanic* case, it may then be difficult to assess whether there really is a gap between the responsibility assigned to the group consisting of everyone involved in the disaster and that assigned to its individual members because it is somewhat unclear how we could measure each agent's causal influence on the disaster. However, in some other cases, it seems clear that there will sometimes be substantial gaps between the responsibility assigned to the group and to its individual members. Consider, for instance, climate change. We know for sure that the climate is changing. According to the Fifth International Panel on Climate Change (IPCC), it is "extremely likely" that "human influence has been the dominant cause of the observed warming since the mid-20th century." The IPCC also points out that

> The atmospheric concentrations of carbon dioxide, methane, and nitrous oxide have increased to levels unprecedented in at least the last 800,000 years. Carbon dioxide concentrations have increased by 40% since pre-industrial times, primarily from fossil fuel emissions and secondarily from net land use change emissions. [. . .] Human influence on the climate system is clear. This is evident from the increasing greenhouse gas concentrations in the atmosphere, positive radiative forcing, observed warming, and understanding of the climate system.[3]

To illustrate the crux of the problem of many hands, we may ask the following: To what extent is each of us morally responsible for global warming? If we believe that, other things being equal, each individual's causal contribution determines his or her moral responsibility, then it seems plausible to conclude that *no single individual* is responsible for global warming. This is because the actions you and I perform have virtually no effect on the global mean temperature. Although it is true that your car emits some small amounts of greenhouse gases, your contribution has virtually no effect on the global mean temperature. The global mean temperature will not change if you leave your car at home. It is true that there will be somewhat fewer molecules of carbon dioxide in the atmosphere, but those extra molecules will not have any (noticeable) effect on the global mean temperature. So although all humans as a *group* are jointly responsible for climate change, it seems that *no single individual* is responsible for it. There is a gap between the responsibility we assign to the group and to its individual members. This shows that global warming is an instance of the problem of many hands (see Figure 12.3).

Before we discuss some possible ways of dealing with the problem of many hands, it is helpful to consider a second, somewhat more precise example discussed in greater detail in chapter 15. Imagine that the lawn in front of the main building of your favorite university has a high aesthetic value. All students and faculty know that the grass is of a very sensitive type, so if too many people walk on it, it will lose its aesthetic value. The options available to each individual student is to either (a) take a shortcut across the lawn and thereby save a few seconds of walking time, or (b) walk around the lawn, which helps preserve its aesthetic value. Thousands of students face this choice every day. Each student can save some time by walking on the grass, and the harm done to the lawn is imperceptible. The grass will look the same no matter whether n or $n + 1$ students walk on it, for every number n.

Figure 12.3

The problem of many hands is a factor to consider in discussions of climate change. According to the IPPC, the concentration of carbon dioxide in the atmosphere has increased by about 40 percent since pre-industrial times, primarily as a result of fossil fuel emissions.   Source: iStock by Getty Images.

Let us imagine that ten thousand students walk on the lawn, and that this completely ruins its aesthetic value. Who is morally responsible for this unacceptable outcome? No *individual member* of the group of ten thousand students had any causal influence on the outcome. The lawn would have looked *exactly the same* no matter whether student *i* had crossed it or not, and this is true for every student *i*. The physical impact each student has on the lawn could, perhaps, be measured in some way. But the physical impact is irrelevant. It is what meets the eye that determines the aesthetic value of the lawn. Therefore, because no single crossing made any evaluatively relevant difference, it would be unreasonable to hold any individual student morally responsible for the outcome in a backward-looking sense (i.e., to ascribe causal or moral liability-responsibility to any individual student). Despite this, it seems clear that the group of ten thousand students *together* harmed the lawn and should be held morally responsible. The upshot is that we, again, face a case in which there is a gap between the responsibility we assign to a group and its individual members.

How we deal with the problem of many hands depends on what we consider to be problematic about it. Consider, for instance, the proposal that we should just accept the fact that there are sometimes gaps between the responsibility we assign to groups and that assigned to individual members of the group. On this view, all of us are responsible for global warming as a group, although no single individual can be held responsible. This proposal is not obviously incoherent if we think the group is a moral agent in its own right. However, if we believe the group is nothing over and above its individual members, and if we believe that claims about guilt and blame should depend on claims about moral responsibility, then it seems to follow that as a member of the group you are blameworthy for global warming—but as a single individual you are not. I leave it to the reader to figure out if it is possible to uphold this view without accepting the contradiction that some individuals are both blameworthy and not blameworthy for global warming.

Another alternative would be to adopt some other principle for distributing responsibility within groups. We will take a closer look at some such principles in the next section.

Therac-25

The Therac-25 was the third generation radiation therapy machine (medical linear accelerator) produced by Atomic Energy of Canada Limited (AECL). Eleven units were installed in American and Canadian hospitals in the early 1980s. Unlike its predecessors, the Therac-6 and Therac-20, the Therac-25 had no hardware interlocks that prevented operators from accidently selecting settings that would be dangerous for the patient. On the new Therac-25, all critical safety features were controlled by software. The operator entered instructions on a keyboard, which were directly translated by the computer into precise instructions for how much and what type of radiation the patient would receive.

On the older models, the Therac-6 and Therac-20, the operator had to manually adjust the settings before the treatment could start. This slowed down the process, which meant fewer patients could be treated per day. Because these machines are extremely expensive, it was desirable to minimize the downtime spent on adjusting settings. This was one of many reasons for developing the faster, but more complex, computer-controlled Therac-25.

On July 26, 1985, a forty-year-old woman received her twenty-fourth radiotherapy treatment from a Therac-25 machine at a cancer clinic in Hamilton, Ontario, Canada. At the time of the treatment, the machine had been in use for over six months. The operator entered the prescribed dose on the keyboard and activated the machine.

However, after about five seconds, the Therac-25 shut down. The display read "treatment pause" and informed the operator that the patient had received "no dose." The operator then decided to try again by pressing the "P" (proceed) key. This was fully in accordance with the instructions in the operating manual. Again, the machine shut down after a couple of seconds. The display read "no dose." The operator pressed the "P" key a third, fourth, and fifth time. Each time the display read "no dose." At that point, the operator asked a hospital service technician to examine the machine. He found nothing wrong with it.

When the treatment ended, the patient complained about a burning sensation, which she described as an "electric tingling shock." After a couple of days, the patient's condition deteriorated. She was hospitalized on July 30 and died of what doctors diagnosed as a massive overdose of radiation on November 3, 1985. AECL, the manufacturer of the Therac-25, was informed of the accident but could not reproduce the error. The company received several reports from other clinics in Canada and the United States about similar problems with the new machine. It was later determined that six patients had received massive overdoses of radiation when undergoing treatments by the Therac-25.

AECL conducted an investigation and did some modifications to the hardware, but they were unable to reproduce the error, so the cause of the problem remained unknown. Two of the reported

(Continued)

accidents had occurred at East Texas Cancer Center in Tyler, Texas. Fritz Hager, a radiotherapy technician working there, was deeply troubled by the harm done by the malfunctioning machine. After days with many unsuccessful attempts, Hager was finally able to reproduce the error. He demonstrated that the overdoses occurred when the operator entered the instruction on the keyboard at *a higher speed* than normal, which caused an overflow error in the computer. Because of this relatively simple bug, the Therac-25 sometimes emitted massive doses of radiation at the same time as the "no dose" text appeared on the display.

The software used in the new machine was written in assembly language. For reasons that remain unclear, all the code was written by a single programmer. He used so-called cargo coding. This means that entire software modules were taken from older versions of the system, which were imported into the Therac-25 software without verifying whether they were fully compatible with the new system. We now know that some of the "cargo" imported from the old system was buggy. These bugs, which caused lethal accidents when used in the Therac-25, also existed in some of the previous versions; but because the Therac-6 and Therac-20 had mechanical safety interlocks, no patients were harmed by them.

The external experts appointed by ACEL found that the software was poorly written. The single programmer responsible for the code had quit his job and could no longer be contacted. The external experts also discovered that the software had not been reviewed by other programmers, and that

it had not been properly tested before the Therac-25 was sold to hospitals. Because the programmer responsible for the software had inserted very few comments and explanations in the code, it was very difficult for external experts to understand the structure of the program.

One of many ethical questions we can ask about this tragic case is the following: Who was morally responsible for the massive overdoses administered by the Therac-25? As noted, the answer depends on whether we are talking about (moral) role-responsibility, causal responsibility, capacity-responsibility, or liability-responsibility. If we interpreted the question as a question about causal responsibility, it may turn out that the Therac-25 case is yet another illustration of the problem of many hands. The single programmer who wrote the buggy software did not cause the death of the patients entirely on his own. The AECL executives could have ensured that the software was reviewed and tested by others. The regulators (the Food and Drug Administration) also had some causal influence on the outcome, as did the hospitals who were eager to switch to a new technology that was not yet mature enough for critical medical applications. It would be naïve to maintain that a single human being was casually responsible for the accidents with the Therac-25.

Discussion question: Imagine that you are the engineer who wrote the buggy software for the Therac-25, which caused numerous fatal burn injuries. What does morality require of you? How can, and should, you remedy your fatal mistake?

MORAL RESPONSIBILTY AND COLLECTIVE OUTCOMES

Because engineers often work in teams, it is important to discuss the allocation of moral responsibility for collective outcomes in some detail. Philosophers Braham and van Hees have proposed three principles for what they call multi-agent interactions with collective outcomes.[4] In this context, the term "multi-agent interactions" just means that the outcome depends on what several agents do. The *Titanic* disaster was the result of multi-agent interactions because no single agent could single-handedly control the disaster. The same holds for climate change and the lawn-crossing example. No single agent can control the global climate or the condition of the lawn in front of the main building of your university.

The following condition characterizes the general structure of multi-agent interactions:

> *Condition 1* (Multi-agent system): There is no single individual who can adopt a course of action that will lead to that individual's chosen outcome irrespective of what others do.

Condition 1 is not a normative condition. It merely describes the structure of the problem discussed here. What all multi-agent interactions have in common is that no single agent can single-handedly control the outcome of his or her chosen course of action.

Braham and van Hees propose two normative conditions for the allocation of moral responsibility for outcomes triggered by multi-agent interactions. Both conditions can be disputed, but taken together they teach us something important about the allocation of moral responsibility for collective outcomes. The first condition states that *some* individual must be responsible for every state of affairs brought about by an action.

> *Condition 2* (Completeness): For every state of affairs, there is at least one individual who is morally responsible for its occurrence.

Although it may seem attractive to maintain that at least one individual should be held responsible for every state of affairs we bring about, this condition is controversial. For instance, it seems possible that humanity as a group, but no single individual, should be held responsible for the negative effects of climate change. If the global climate changes catastrophically in the next few decades, it seems that all of us together, but no single individual, should be held responsible for this. Let us nevertheless accept Condition 2, for the sake of the argument.

The next condition states that responsibility allocations should not be fragmented. Braham and van Hees explain that "fragmented responsibility occurs when a combination of actions by different individuals leads to an outcome for which at least some of the responsible individuals are responsible for different features of the outcome."[5] This sounds more abstract than it is. To illustrate the notion of fragmentation, imagine that two gangsters shoot at an innocent victim through the *left* and *right* windows of his car. Both gangsters destroy one window each, but only the bullet fired by the first gangster hits and kills the victim. If we believe that the moral responsibility for this complex outcome should be fragmented, then we could say that the second gangster is responsible for destroying one of the windows but not for killing the victim, while the first gangster is responsible for the damage done to the other window and for the death of the victim. However, according to Braham and van Hees, such fragmentation of responsibility is often undesirable. From a moral point, both gangsters were equally responsible for killing the victim and destroying the car. The second gangster cannot deny his responsibility by arguing that the other gangster contributed more to the outcome.

> *Condition 3* (No fragmentation): There is no fragmentation of moral responsibility: All individuals who are morally responsible for some feature of an outcome are morally responsible for the very same features.

What makes these three conditions interesting is the fact that every principle for allocating moral responsibility that satisfies Conditions 2 and 3 violate Condition 1, meaning that no principle can satisfy all three. We can accept any two of the conditions, but not all of them. The proof for this formal theorem depends on some technical assumptions about what counts as an "outcome" and a "state of affairs," which are widely accepted by ethicists working with formal methods.

The fact that not all three conditions can be satisfied simultaneously does not entail that it is impossible to allocate moral responsibility for collective outcomes in a coherent manner. The point is just that we face a theoretical choice. We can either give up the idea that at least one *individual* must be responsible for the occurrence or every state of affairs, or we can accept *fragmentations* in responsibility allocations. I leave it to the reader to figure out which option is most attractive. The most important lesson of this discussion is methodological. Rather than just making some controversial claims about who is, or is not, morally responsible for a certain type of outcome, Braham and van Hees teach us that we can use formal methods for analyzing abstract principles for responsibility allocations in a very precise and technically illuminating manner.

CASE 12-4

Who Is Responsible for Tech Addiction?

Stanford Students Against Addictive Devices (SSAAD) is an organization founded by computer science students at Stanford. The group claims that 69 percent of adults check their iPhone hourly; and they cite scientific studies indicating that "iPhone addiction" causes stress, harms relationships, and undermines productivity at work. According to SSAAD, the maker of the iPhone, Apple Inc., should assume responsibility for helping people curb their dependence by modifying the software such that their devices become less addictive. Similar ideas have been put forward by the journalist Farhad Manjoo. Here is an excerpt from an article published in the *New York Times* in January 2018:

> Tech "addiction" is a topic of rising national concern. [...] I got to thinking about Apple's responsibility last week when two large investors wrote an open letter asking the company to do more about its products' effects on children. I was initially inclined to dismiss the letter as a publicity stunt; if you're worried about children and tech, why not go after Facebook? But when I called several experts, I found they agreed with the investors. Sure, they said, Apple isn't responsible for the excesses of the digital ad business,

but it does have a moral responsibility to—and a business interest in—the well-being of its customers. And there's another, more important reason for Apple to take on tech addiction: because it would probably do an elegant job of addressing the problem. [...]

Imagine if, once a week, your phone gave you a report on how you spent your time, similar to how your activity tracker tells you how sedentary you were last week. It could also needle you: "Farhad, you spent half your week scrolling through Twitter. Do you really feel proud of that?" It could offer to help: "If I notice you spending too much time on Snapchat next week, would you like me to remind you?"

Another idea is to let you impose more fine-grained controls over notifications. Today, when you let an app send you mobile alerts, it's usually an all-or-nothing proposition—you say yes to letting it buzz you, and suddenly it's buzzing you all the time. [...] Apple could set rules for what kind of notifications were allowed. [...] Apple released a statement last week saying it cared deeply "about how our products are used and the impact they have on users and the people around them," adding that it had a few features on addiction in the works. Apple hardly ever talks about future products, so it declined to elaborate on any of its ideas when I called. Let's hope it's working on something grand.

Discussion question: Do you agree with Mr. Manjoo that Apple is responsible for its products' effects on children? Or are the parents who let their children use the addictive products solely responsible? Give reasons for and against.

REVIEW QUESTIONS

1. Is a person morally responsible for her wrongdoing when acting under the influence of alcohol? How did Aristotle analyze this question?
2. Explain H. L. A. Hart's notions of role-responsibility, causal responsibility, capacity-responsibility, and liability-responsibility.
3. Discuss the distinction between backward-looking and forward-looking responsibility. Is this a useful distinction?
4. What is the problem of many hands, and why is the *Titanic* disaster a good illustration of this? Who was responsible for the *Titanic* disaster?
5. According to the IPCC, "The atmospheric concentrations of carbon dioxide, methane, and nitrous oxide have increased to levels unprecedented in at least the last 800,000 years." Who is responsible for this problem? Apply the concepts discussed in this chapter.
6. Explain the idea that there might sometimes be a gap between the responsibility we assign to the group and to its individual members.
7. What is a multi-agent system and why are such systems important for discussions of moral responsibility?

REFERENCES AND FURTHER READINGS

Braham, M., and M. van Hees. 2018. "Voids or Fragmentation: Moral Responsibility for Collective Outcomes." *Economic Journal* 128, no. 612: F95–F113.

Gillum, Jack D. 2000. "The Engineer of Record and Design Responsibility." *Journal of Performance of Constructed Facilities* 14, no. 2: 67–70.

Hart, H. L. A. 2008. *Punishment and Responsibility: Essays in the Philosophy of Law.* Oxford, UK: Oxford University Press.

Intergovernmental Panel on Climate Change. 2014. "Climate Change 2014: Mitigation of Climate Change." Geneva, Switzerland: IPCC.

Manjoo, Farad. January 17, 2018. "It's Time for Apple to Build a Less Addictive iPhone." *New York Times.*

Marcus, G. J. 1976. *The Maiden Voyage.* London: New English Library.

McGinn, R. 2018. *The Ethical Engineer: Contemporary Concepts and Cases.* Princeton, NJ: Princeton University Press.

Pachauri, R. K., M. R. Allen, V. R. Barros, J. Broome, W. Cramer, R. Christ, J. A. Church, et al. 2014. *Climate Change 2014: Synthesis Report. Contribution of Working Groups I, II and III to the Fifth Assessment Report of the Intergovernmental Panel on Climate Change.* Geneva, Switzerland: IPCC.

van de Poel, I., J. N. Fahlquist, N. Doorn, S. Zwart, and L. Royakkers. 2012. "The Problem of Many Hands: Climate Change as an Example." *Science and Engineering Ethics* 18, no. 1 (2012): 49–67.

Ross, S. W. D. 1954. *The Nicomachean Ethics of Aristotle.* Translated and introduced by Sir D. Ross. Oxford, England: Oxford University Press.

NOTES

1. R. McGinn, *The Ethical Engineer: Contemporary Concepts and Cases* (Princeton, NJ: Princeton University Press, 2018), 194.

2. H. L. A. Hart, *Punishment and Responsibility: Essays in the Philosophy of Law* (1968; repr., Oxford, England: Oxford University Press, 2008), 211.

3. Intergovernmental Panel on Climate Change, "Climate Change 2014: Mitigation of Climate Change" (Geneva, Switzerland: IPCC, 2014).

4. M. Braham and M. van Hees, "Voids or Fragmentation: Moral Responsibility for Collective Outcomes," *Economic Journal* (forthcoming).

5. Ibid.

Engineering and Society

CHAPTER 13

Technology Assessments and Social Experiments

New technologies often have unforeseen effects on society. Some are beneficial, but others are harmful in ways that are difficult to predict. Consider cell phones equipped with high-resolution cameras. While often handy, we never foresaw when introduced that they might be used for filming sexual assaults, or for paparazzi photos of celebrities in restaurants. The introduction of technology to society deserves attention from ethicists. Should new technologies be systematically assessed and approved before being introduced? The FDA takes great care to ensure that new drugs and medical treatments undergo extensive clinical trials before they are made available to patients. No similar system exists for ordinary, non-medical technologies. Perhaps *all* new technologies should be monitored and assessed as carefully as medical products.

In this chapter, we consider two proposals for how to monitor and assess new technologies. The first is to formally assess technology prior to its introduction to society as practiced in the United States between 1972 and 1995. This assessment may include discussions of the ethical issues raised by the technology. The second proposal is to treat new technologies as ongoing *social experiments* that require continuous monitoring and assessment, even after introduction. This idea has never been formally tested, at least not in an institutionalized form, but several philosophers have suggested that this is a promising proposal.

TECHNOLOGY ASSESSMENT

The Office of Technology Assessment (OTA) was established by the US Congress in 1972. After a heated political debate, it was eventually defunded and abolished in 1995. During those twenty-three years, its mission was to "provide early indications of the probable beneficial and adverse impacts of the applications of technology." In the bill passed by Congress to create the OTA, it was pointed out that "it is essential that, to the fullest extent possible, the consequences of technological applications be anticipated, understood, and considered in determination of public policy on existing and emerging national problems." Today, technology assessments are still being performed in Germany (at the Institute of Technology Assessment, ITAS, in Karlsruhe), the Netherlands (at the Rathenau Institute in The Hague), as well at academic centers in Switzerland, Austria, the UK, and other European countries.

Over its twenty-three years of existence, the OTA produced more than 700 reports. They had titles such as "Affordable Spacecraft: Design and Launch Alternatives" (1990), "Transportation of Liquefied Natural Gas" (1977), and "The Market for Wheelchairs: Innovations and Federal Policy" (1984). In an influential report entitled "High-performance Computing and Networking for Science," published in September 1989, the OTA accurately predicted several key features of the Internet as we know it today, many years before it became widely used in universities and research labs around the world. Here is a representative excerpt:

> Within the next decade, the desks and laboratory benches of most scientists and engineers will be entry points to a complex electronic web of information technologies, resources and information services, connected together by high-speed data communication networks. These technologies will be critical to pursuing research in most fields. Through powerful workstation computers on their desks, researchers will access a wide variety of resources, such as:
>
> - an interconnected assortment of local campus, State and regional, national, and even international data communication networks that link users worldwide;
> - specialized and general-purpose computers including supercomputers, mini-supercomputers, mainframes, and a wide variety of special architectures tailored to specific applications;
> - collections of application programs and software tools to help users find, modify, or develop programs to support their research;
> - archival storage systems that contain specialized research databases;
> - experimental apparatus—such as telescopes, environmental monitoring devices, seismographs, and so on—designed to be set-up and operated remotely;
> - services that support scientific communication, including electronic mail, computer conferencing systems, bulletin boards, and electronic journals;
> - a "digital library" containing reference material, books, journals, pictures, sound recordings, films, software, and other types of information in electronic form;
> - and specialized output facilities for displaying the results of experiments or calculations in more readily understandable and visualizable ways.[1]

Every technological advancement predicted by the OTA in this report is available today. Some features took more than a decade to develop, and the Internet is no longer exclusively used by researchers; but on the whole, the report was strikingly accurate. Academics began to use the Internet for sending emails on a large scale in the early 1990s, but the digital libraries containing books, films, and sound recordings predicted in the report became available about a decade later. All 700 reports published by the

OTA can be accessed online for free, including the very report that predicted the existence of such a digital library.

Although the quality of the OTA reports was generally high, social and ethical implications of new technologies seem to have been much more difficult to predict than purely technical ones. For instance, in its report on the Internet, the OTA included a section entitled "Longer Term Science Policy Issues" in which the authors argued that "equity of access to scientific resources" could become an issue in the future, as well as "legal issues, data privacy, ownership of data, copyright." While true, this was not very helpful. The really important social and ethical implications of the Internet were totally lacking in the report. The OTA failed to predict how the Internet would revolutionize society far beyond universities. Instead of being a tool for researchers, the Internet as we know it today is used by everyone at all hours of the day. It has drastically altered the business model of travel agencies, banks, and newspapers; and many of us spend much of our Internet time conversing with friends and foes on social media.

One conclusion from all this is that the uncertainties of a technology assessment are so great that the morally relevant information we get from them tends to be of somewhat limited practical value. The OTA report on the Internet was not very useful for decision makers who wanted to understand the social and ethical implications of this new technology. An underlying reason for this, which is also a general reason for being somewhat skeptical about the value of technology assessments in engineering ethics, is that social and moral issues are often not robust in the following sense: small changes to the way a technology is used or designed often have large implications for the social and moral issues we are interested in. For instance, in the OTA assessment of the Internet, they overlooked the fact that everyone would be using the Internet, not just researchers, which turned out to make a huge difference from a social and moral point of view.

NEW TECHNOLOGIES AS SOCIAL EXPERIMENTS

A technology assessment is a *static* evaluation of a new technology performed at a single point in time. However, the way technologies are developed and used sometimes change, meaning that a static technology assessment may quickly become obsolete. It has been proposed that the best way to address this issue is to think of new technologies as ongoing *social experiments* that need to be *continuously* monitored and assessed, just like research experiments. The OTA acknowledged the analogy between new technologies and research experiments in its report on the Internet: "the current Internet is, to an extent, an experiment in progress, similar to the early days of the telephone system. Technologies, uses, and potential markets for network services are still nascent. Patterns of use are still evolving; and a reliable network has reached barely half of the research community."[2]

The suggestion to think of new technologies as social experiments was introduced by engineer Roland Schinzinger in 1973. The idea gradually received more attention in the 1980s as Schinzinger and philosopher Mike W. Martin co-authored an influential textbook on engineering ethics in which the technology-as-a-social-experiment metaphor was extensively discussed.[3] If we were to follow their proposal, that would have important implications for the ethical evaluation of new technologies. Rather than asking ourselves if some technology X is ethically acceptable, which is a question we know to be difficult to answer, we could replace that question with a new one, which might be easier to answer: "Is technology X an ethically acceptable social experiment?"

There is a significant body of literature that addresses the ethics of scientific experimentation. The Nuremberg Code is a particularly influential document, in which the principle of informed consent plays a pivotal role. (See Case 13-1).

However, the principle of informed consent emphasized in the Nuremberg Code may be difficult to apply to engineering contexts, as noted in chapter 10. It is often impractical or impossible to ask *all* users of a new technology for informed consent before the technology is introduced. Imagine, for instance, that General Electric develops a new and more efficient engine for commercial, wide-body aircrafts. Airlines using the new engine can hardly be required to inform passengers about the pros and cons of the new engine and ask them to sign consent forms before boarding. Another example is the Internet. The author of this textbook never gave his informed consent to the introduction of the Internet in society. Moreover, even if someone had bothered to ask me, I would not have been able to make an *informed* decision. It seems clear that the principle of informed consent, as it is used in the biomedical domain, is not always applicable to decisions concerning new technologies.

CASE 13-1

The Nuremberg Code for Scientific Experimentation

The Nuremberg Code is respected by thousands of researchers around the world. It was drafted by the Nuremberg Tribunal after World War II in response to the ethically unacceptable experiments performed by Nazi doctors on prisoners in concentration camps. The code mentions several principles that could be applied to new technologies:

1. The voluntary consent of the human subject is absolutely essential. This means that the person involved should have legal capacity to give consent; should be so situated as to be able to exercise free power of choice, without the intervention of any element of force, fraud, deceit, duress, overreaching, or other ulterior form of constraint or coercion; and should have sufficient knowledge and comprehension of the elements of the subject matter involved as to enable him to make an understanding and enlightened decision.

2. The experiment should be such as to yield fruitful results for the good of society, unprocurable by other methods or means of study, and not random and unnecessary in nature.

3. The degree of risk to be taken should never exceed that determined by the humanitarian importance of the problem to be solved by the experiment.

4. Proper preparations should be made and adequate facilities provided to protect the experimental subject against even remote possibilities of injury, disability or death.

5. During the course of the experiment the human subject should be at liberty to bring the experiment to an end if he has reached the physical or mental state where continuation of the experiment seems to him to be impossible.[4]

Discussion question: Apply the five principles of the Nuremberg Code to self-driving cars. Are all, or some, of the five principles applicable to this technology?

Martin and Schinzinger propose that we can avoid this problem by weakening the principle of informed consent. In their view, it is sufficient that the information a *rational person* would need for making a decision about the new technology has been *widely disseminated*. Philosopher of technology Ibo van de Poel has objected to this, saying that the relevant information a rational person would need when deciding to consent to a new technology is in many cases unavailable. If asked to consent to a treatment in the hospital, your doctor can tell you roughly how much pain you can expect to feel. But what information could have been "widely disseminated" about the pros and cons of the Internet in 1989 that would have helped a rational person make an informed decision about this technology?

Some of the other principles of the Nuremberg Code (Case 13-1) might be easier to apply. If new technologies are conceived as ongoing social experiments, we will sometimes be able to stop experiments that do not "yield fruitful results for the good of society" (principle 2). Computer viruses are instances of a technology that does not yield fruitful results for the good of society, nor does your favorite computer game.

While principles 3 and 4 seem to apply to many new technologies, principle 5 is in many cases difficult to satisfy. Once we have started to use a technology, it can be very difficult to stop. If I, for instance, decide that I would like to stop using emails because I have "reached the . . . mental state where continuation of the experiment seems . . . impossible," because this takes up too much of my time, I would not know how to quit. Everyone at work tells me to read and respond to emails, so I could not stop doing so without being fired. However, this hardly shows that it is morally wrong to send emails.

Van de Poel seeks to remedy these and other problems with the Nuremberg Code by proposing other criteria for evaluating new technologies framed as social experiments. According to van de Poel, the acceptability of a new technology, conceived as an ongoing social experiment, depends on whether the following criteria are met:[5]

1. Absence of other reasonable means for gaining knowledge about risks and benefits
2. Monitoring of data and risks while addressing privacy concerns
3. Possibility and willingness to adapt or stop the experiment
4. Containment of risks as far as reasonably possible
5. Consciously scaling up to avoid large-scale harm and to improve learning
6. Flexible set-up of the experiment and avoidance of lock-in of the technology
7. Avoid experiments that undermine resilience
8. Reasonable to expect social benefits from the experiment
9. Clear distribution of responsibilities for setting up, carrying out, monitoring, evaluating, adapting, and stopping of the experiment
10. Experimental subjects are informed
11. The experiment is approved by democratically legitimized bodies
12. Experimental subjects can influence the setting up, carrying out, monitoring, evaluating, adapting, and stopping of the experiment
13. Experimental subjects can withdraw from the experiment

14. Vulnerable experimental subjects are either not subject to the experiment or are additionally protected or particularly profit from the experimental technology (or a combination)
15. A fair distribution of potential hazards and benefits
16. Reversibility of harm or, if impossible, compensation of harm

To what extent these conditions can be applied to new technologies may vary from case to case, as illustrated in the Autonomous Military Robots case (Case 13-2).

CASE 13-2

Autonomous Military Robots

Military labs in the United States and elsewhere are currently developing small autonomous drones for warfare. Stuart Russell, a professor of computer science at UC Berkeley, calls these military robots "slaughterbots." He envisions a dystopic future in which swarms of slaughterbots are instructed to attack technologically inferior enemy armies, perhaps with instructions to kill everyone wearing a uniform, or—if the drones are equipped with face recognition technology—specific individuals, such as members of a certain ethnic group. The advantage of autonomous drones is that the attacker's soldiers will never be exposed to any risk. Wars can be fought by civilians working in ordinary offices thousands of miles away.

However, terrorists and wealthy criminals may also be tempted to use slaughterbots. If we can kill our enemies by using small flying robots (see Figure 13.1) equipped with face recognition technology, anyone can become a mass murder at any point in time. Terrorists and racists could kill their adversaries with great precision; and if the robots are mass produced, they could wipe out entire cities without destroying buildings or other valuable assets.

Are autonomous military robots unethical? We can think of them as an *ongoing social experiment.* We can then apply (some of) the criteria proposed by van de Poel; to start with, criterion 3 on van de Poel's list would not be satisfied because it would probably be impossible to stop or modify the experiment (the slaughterbots) once they become available. The same goes for condition 9; there is no clear distribution of responsibilities for setting up, carrying out, monitoring, evaluating, adapting, or stopping the experiment. Criteria 10 and 11 are

Figure 13.1
The Sperwer Epervier D-47 was one of the first military drones developed at the request of NATO. It is currently on display in a museum in Brussels.
Source: Wikimedia.

also problematic, as are criteria 12–16, for obvious reasons. It seems clear that if autonomous military robots (slaughterbots) are conceptualized as ongoing social experiments, then they would not qualify as a morally acceptable experiment.

Discussion question: Suppose we agree that autonomous military robots are unacceptable social experiments, for the reasons just mentioned. If so, what could engineers do to stop this technology from being developed? Can you think of any other technology we have chosen not to develop for ethical reasons? Or is it impossible to stop new technologies from being introduced in society? (Cf. the discussion of technological determinism in chapter 3.)

SOME CRITICAL QUESTIONS

The suggestion that we should think of new technologies as social experiments is controversial. This is partly because it might seem overly optimistic to think that it would be easier to adjudicate whether a social experiment is ethically acceptable than it is to adjudicate whether a new technology is ethically acceptable. The new question ("Is technology X an ethically acceptable social experiment?") is often no easier to answer than the old one ("Is technology X an ethically acceptable technology?"). Our understanding of research ethics is not much better than our understanding of ethics in general. So, what would the point be of replacing a difficult question by another equally difficult question?

The traditional debate between ethicists defending consequentialist and deontological theories pops up in research ethics as well as in many other areas of applied ethics. Conflicts between different ethical theories cannot be easily solved or avoided. The debate over embryonic stem cell research is a good illustration of this. Some argue that stem cell therapies might lead to good consequences for a large number of people, which they think is more important than deontological concerns about the moral status of the embryo. Others take the opposite view. How do we decide which camp is right? It seems that it is unhelpful to frame this question as a problem of research ethics. The ethical conflict is still present, unresolved, and open for everyone to see.

Another, more fundamental objection is that a good answer to a question about the acceptability of a social experiment may not be a good answer to the original question of whether we should accept the new technology. The two questions do not address the same issue. Therefore, the original question about the acceptability of the technology would still have to be answered, even if we would be able to answer the new question about the acceptability of the social experiment. In some cases, a new technology should be accepted (and used) even if it would *not* pass as an acceptable social experiment. The reason is that other considerations, having nothing to do with research ethics, can be relevant for deciding whether a new technology should be accepted or not.

President Truman's decision to drop two nuclear bombs in Japan in August 1945 might serve as a good illustration (see Figure 13.2). If we were to evaluate Truman's decision from a research ethical point of view, there is no doubt that a large number of considerations that are typically considered to be relevant in discussions of research ethics were violated in this case, such as "ask for informed consent"; "don't inflict harm on innocent people against their will"; and so forth. But this does not show that Truman's decision to drop the nuclear bombs was ethically unacceptable. In 1945, the United States was at war with Japan, and Truman did what he had good reason to think would be the best way to end the war. Some historians have argued that his decision might have saved the lives of millions of innocent people. To think of the decision to use nuclear weapons against Japan from a research ethical point of view would not have been of any help for Truman. The decision he was facing, whether to use a new and potentially very dangerous technology for warfare, could be answered only by figuring out whether it was ethically acceptable to use this new technology for that purpose. There was no other question he could have asked that would have been easier but yet meaningful to answer.

What could be said in response to these objections? It seems that the best response might be to weaken the claim a bit. Instead of claiming that questions about the ethical

Figure 13.2
In August 1945, President Truman decided to drop two nuclear bombs in Japan. This photo, taken by Charles Levy, shows the bomb dropped over Nagasaki on August 9, 1945. Was this an ethically acceptable social experiment?   Source: National Archives image (208-N-43888).

acceptability of social experiments should *replace* questions about the ethical acceptability of technologies, one could, in a more cautious manner, argue that the new question adds new, valuable perspectives to the ethical debate, which would otherwise have been overlooked. So although the old question still has to be answered, and is likely to remain difficult to answer, we might discover some new (equally difficult questions) that may be relevant and that would otherwise have been neglected. By shifting the focus of the debate, we may not be able to solve the original problem, but we achieve other things that could be valuable. We might, for instance, discover that a question we knew to be difficult to answer is even more difficult to answer than we thought because additional ethical perspectives need to be taken into account.

Another possible response could be to argue that if a technology passes the research ethical test (or whatever test would be triggered by the new question), then it is unlikely to be unacceptable because of some other reason. In contemporary research ethics, nearly no ethical worry seems to be too insignificant to warrant concern. There is probably no other area in society in which ethical issues are taken so seriously. Therefore, we can use the research ethical test for identifying which new technologies need to be carefully monitored. The technology-as-a-social-experiment approach would then be a useful heuristic, which we might have good pragmatic reasons for adopting.

A counterargument to this type of response is that if we weaken the claim and no longer insist that ethical questions about new technologies can be replaced by questions about social experiments, then the technology-as-a-social-experiment metaphor loses nearly all of its intellectual content. If we were to admit that the original question we set out to tackle still needs to be answered ("Is technology X ethically acceptable"), that is, that the new question cannot replace the old one, then the old question still remains to be answered. By bringing up various pragmatic reasons for discussing the new question, we do not solve the original intellectual problem: Should the new technology be accepted or not? The technology-as-a-social-experiment metaphor would be worth serious interest only if it could, literally speaking, *replace* the old and difficult question about whether a new technology should be accepted.

REVIEW QUESTIONS

1. What is a technology assessment? What question or questions does a technology assessment seek to answer?
2. The Office of Technology Assessment (OTA) predicted in 1989 that "within the next decade . . . a 'digital library' containing reference material, books, journals, pictures, sound recordings, films, software, and other types of information in electronic form" would become available. How were they able to make that prediction before these innovations were actually made? Does the accuracy of this prediction support technological determinism (see Chapter 3)?
3. Discuss some of the problems with using technology assessments for making moral evaluations of new technologies.
4. How should we understand the suggestion that it is fruitful to think of new technologies as social experiments?
5. Discuss what role the principle of informed consent should, or should not, play if we are to think of new technologies are social experiments.
6. Can questions about the ethical acceptability of social experiments *replace* questions about the ethical acceptability of new technologies? If not, are questions about the ethical acceptability of social experiments important for some other reason?
7. Propose at least one other method for evaluating the ethical acceptability of a new technology before it is introduced in society.

REFERENCES AND FURTHER READINGS

Martin, M. W., and R. Schinzinger. 1983. *Ethics in Engineering*. New York, McGraw-Hill Books Company.

"Nuremberg Code." 1949. In A. Mitscherlich and F. Mielke, *Doctors of Infamy: The Story of the Nazi Medical Crimes*, xxiii-xxv. New York: Schuman.

Peterson, M. 2017. "What Is the Point of Thinking of New Technologies as Social Experiments?" *Ethics, Policy & Environment*, no. 20: 78–83.

van de Poel, I. 2016. "An Ethical Framework for Evaluating Experimental Technology." *Science and Engineering Ethics*, no. 22: 667–686.

Shuster, Evelyne. 1997. "Fifty Years Later: The Significance of the Nuremberg Code." *New England Journal of Medicine* 337, no. 20: 1436–1440.

US Congress, Office of Technology Assessment. September 1989 *High Performance Computing and Networking for Science—Background Paper. OTA-BP-CIT-59*. Washington, DC: US GPO.

NOTES

1. US Congress, Office of Technology Assessment, *High Performance Computing and Networking for Science—Background Paper, OTA-BP-CIT-59* (Washington, DC: US GPO, September 1989), 1.

2. Ibid. 28–29.

3. M. W. Martin and R. Schinzinger, *Ethics in Engineering* (New York, McGraw-Hill Books Company, 1983).

4. "The Nuremberg Code," in A. Mitscherlich and F. Mielke, *Doctors of Infamy: The Story of the Nazi Medical Crimes* (New York: Schuman, 1949), xxiii–xxv. See also E. Shuster, "Fifty Years Later: The Significance of the Nuremberg Code," *New England Journal of Medicine* 337, no. 20 (1997): 1436–1440.

5. I. van de Poel, "An Ethical Framework for Evaluating Experimental Technology," *Science and Engineering Ethics,* no. 22 (2016): 680.

CHAPTER 14

A Critical Attitude to Technolgy

In the 1960s and 1970s, several philosophers of technology pointed out that not all new technologies make the world better. Some modern technologies enable us to change the world in unprecedented ways, or even extinguish life on the planet as we know it. If there is no plausible scenario in which it is morally permissible to, say, use atom bombs or other weapons of mass destruction, then the mere existence of those technologies seems to be morally problematic.

According to *technological pessimists*, the overall value of technological progress is questionable (see chapter 1). The atom bomb and other weapons of mass destruction are important examples, but the concern raised by these thinkers is more general. Modern technology has enabled us to communicate via cell phones, watch movies at home, and travel to almost any place on the planet. But has any of these technologies made our lives better? Apart from medical technologies that help us overcome real obstacles in life, it seems that many modern technologies actually do relatively little to improve our quality of life. What ultimately matters to us is not the number of cell phones we own but social interactions in the real world and other nontechnological aspects of life.

In this chapter, we shall take a closer look at technological pessimism and contrast it with *technological optimism*. We will also consider a third, intermediate, position, according to which we should approach new technologies with a *critical attitude*. On this intermediate view, technological pessimists are too cynical and gloomy, while optimists are naïve. Although it is true that some technologies sometimes have negative effects on society, this is not true of all, or even most, technologies. For instance, numerous medical technologies have improved the lives of billions of people around the globe. We should be grateful for all the important work engineers do, rather than blame them for occasional mishaps. To use a metaphor proposed by my colleague Dr. Glen Miller, we should operate in much the same way as restaurant critics: "The starter was a bit too salty, but the entrée was magnificent, although the wine suggested by the sommelier was not a perfect match." If we adopt this critical attitude to technology, we should make negative remarks only when appropriate, typically at an early stage when the technology is still in the design phase. We have no reason to make negative or positive remarks about technology *as such*. Some particular technologies are good, and others are bad, just as some particular restaurants are good, and others are bad. It would be absurd to say that restaurants *as such* are good or bad.

CASE 14-1

Self-Driving Cars and Technological Optimism

On May 7, 2016, a Tesla Model S crashed into the trailer of a big-rig truck on a highway in Florida. The National Transportation Safety Board (NTSB) concluded that the vehicle's autopilot feature had been engaged at the time of the crash. The Tesla failed to identify the truck and apply the brakes. This made the car crash into the trailer at a speed of 74 miles per hour. The owner of the Tesla, Joshua Brown, died immediately.

Mr. Brown was first person to die in an accident with a self-driving car. According to newspaper accounts, he was a technological optimist: he emphasized the positive aspects of new technologies in general, including his Tesla. At the time of the crash, he had driven his vehicle more than forty-five thousand miles in less than nine months. Before the accident, he also uploaded several YouTube videos showing him cruising along with the autopilot feature engaged. In a video called "Autopilot Saves Model S," Mr. Brown demonstrates how the Tesla's autopilot avoided an accident by swerving to the right as another vehicle approached from the left.

In its report, the NTSB pointed out that Mr. Brown overestimated the abilities of the autopilot by not following the instruction in the owner's manual to never let the car drive itself without human supervision. According to the NTSB, Mr. Brown's "pattern of use of the Autopilot system indicated an over-reliance on the automation and a lack of understanding of the system limitations." Despite this, the NTSB concluded that Tesla bears some of the blame because the autopilot had been designed such that it could easily be misused: "If automated vehicle control systems do not automatically restrict their own operation to conditions for which they were designed and are appropriate, the risk of driver misuse remains." Tesla modified the software after the accident such that the autopilot feature now works only if the driver keeps his or her hands on the wheel and is ready to disengage the autopilot (see Figure 14.1).

Needless to say, it may not be trivial from a technical point to design autonomous vehicles that eliminate or significantly reduce the risk of accidents. However, everyone agrees that doing so would be morally desirable. If autonomous vehicles can significantly reduce the number of deaths on the highways, then that is a strong reason for using self-driving cars. About thirty-seven thousand people die in traffic accidents in the United States every year, many of which are caused by human errors.

Unfortunately, not all ethical issues concerning autonomous systems are that easy. Should, for instance, an autonomous vehicle be instructed to protect pedestrians on the sidewalk to the same extent as its passengers? (At least one major automaker, Mercedes-Benz, has declared that its self-driving cars will prioritize occupant safety over pedestrians.[1]) Another, somewhat more visionary, ethical issue designers of autonomous systems may face in the coming years is the following: if tomorrow's super-smart robots and other future autonomous systems become smarter than us, they may decide to maximize their own

Figure 14.1

Mr. Brown's Tesla S after the fatal crash.
Source: Florida Highway Patrol investigators.

"well-being" at our expense. To put it crudely, these autonomous systems may decide—if they are truly super-smart and fully autonomous—to work for their own benefit rather than ours. Is this something we should tolerate, or should autonomous systems be programmed to always promote human well-being?

Stuart Russell, a computer scientist at Berkley, has suggested that we should design autonomous systems with values that "are aligned with those of the human race." He calls this *the value alignment problem*. As he sees it, the challenge we face is to build autonomous systems with ethical priorities that do not pose a threat to human beings. He claims that "the machine's purpose must be to maximize the realization of human values."[2]

The IEEE, the world's largest professional organization for engineers, has issued a report on the value alignment problem entitled *Ethically Aligned Design: A Vision for Prioritizing Human Well-being with Autonomous and Intelligent Systems* (A/IS).[3] This document is part of the *IEEE Global Initiative on Ethics of Autonomous and Intelligent Systems*. The authors claim that "A/IS should be designed and operated in a way that both respects and fulfills human rights, freedoms, human dignity, and cultural diversity." The authors also note that autonomous systems "should always be subordinate to human judgments and control. [. . .] If machines engage in human communities as autonomous agents, then those agents will be expected to follow the community's social and moral norms."

These ethical concerns seem to indicate that a critical attitude to autonomous systems and other new technologies could, at least sometimes, be warranted. The outright technological optimism displayed by some drivers of self-driving cars may lead to unforeseen and unwanted negative consequences, as in the Tesla crash in 2016. New technologies often make the world better in certain respects; but to prevent future accidents, it seems wise to continue to critically monitor the introduction of new technologies in society.

Discussion question: What are, in your opinion, the three most important ethical issues triggered by the introduction of self-driving cars in society? Can these issues be addressed (or even eliminated) by the modifying the technology used in self-driving cars? Or are new regulations needed?

THE IMPERATIVE OF RESPONSIBILITY

The German-American philosopher Hans Jonas (1903–1993) was one of the most important critics of modern technology in the twentieth century. He argued that "the altered nature of human action, with the magnitude and novelty of its works and their impact on man's global future, raises moral issues for which past ethics . . . has left us unprepared."[4] According to Jonas, we therefore need to develop a new ethical principle: the imperative of responsibility.

Let us unpack Jonas's complex ideas word by word. To start with, Jonas thinks that modern technologies have *altered the nature* of human action. While it is true that we still perform actions by moving our bodily parts in much the same way as our ancestors did, the actions we perform today are fundamentally different. The range of alternatives available to us is much greater, and some of the potential consequences are much more severe. For these reasons, Jonas believes that "the human condition" (the essential components and characteristics of our existence) is no longer fixed once and for all. Unlike previous generations, members of the current generation can change the human condition. This can impact man's global future. Previous generations could not do anything that jeopardized the existence of the entire species, but we can. This is a radical difference compared to how it used to be.

What makes Jonas's remarks interesting is partly his claim that no *past ethics* (i.e., no traditional moral theory or principle) can do justice to the moral issues triggered by modern technology. This is a controversial claim. Utilitarians would argue that the traditional utilitarian criterion of moral rightness applies universally, no matter whether our actions impact man's global future. Jonas's response to this is that utilitarianism and other traditional theories are *anthropocentric* (concerned only with humans), which makes them inapplicable to ethical issues that affect the future of the planet. A possible response could be that it would be easy to extend the utilitarian theory (and, arguably, other traditional theories too) to whatever additional concerns the recent changes to the human condition gives us reason to consider. In fact, Bentham and other influential utilitarians have repeatedly stressed the importance of including animal well-being and the interests of other sentient beings in the utilitarian calculation; so in that sense, it is a mistake to claim that utilitarianism is an anthropocentric theory.

Jonas puts forward his own ethical principle as an improvement of Kant's universalization test (see chapter 6). Like Kant, he expresses his principle, the imperative of responsibility, through multiple formulations that are meant to be equivalent. Here are two of the clearest statements:

1. Do not compromise the conditions for an indefinite continuation of humanity on earth.[5]
2. Act so that the effects of your action are compatible with the permanence of genuine human life.[6]

Jonas admits that it is not contradictory in any logical or practical sense to violate these principles. It is, for instance, not logically or practically impossible for the present generation to compromise the conditions for an indefinite continuation of humanity of earth. If we decide to use vast amounts of natural resources today but leave nothing to future generations, then we are not *contradicting* ourselves. Therefore, Jonas's imperative of responsibility cannot be justified in the same way as Kant's categorical imperative. Somewhat surprisingly, Jonas's argument for his principle is instead based on the following idea:

> The ethical axiom which validates the rule is therefore as follows: Never must the existence of man as a whole be made a stake in the hazards of action. It follows directly that bare possibilities of the designated order are to be regarded as unacceptable risks which no opposing possibilities can render more acceptable . . . [a] "go-for-broke" calculation of risk, objectionable also in other respects, is in error already by . . . the risk of *infinite* loss.[7]

Put in plain English, Jonas argues that the finite benefits we can get from the use of modern technology, no matter how probable, will always be outweighed by the tiny probability of infinite loss. If we all go extinct in a nuclear war, or if our planet faces catastrophic environmental degradation, then the potential benefits of modern technology no longer matter.

Is this argument convincing? The weakest part seems to be the claim that we should assign *infinite* negative value to the extinction of humanity. Modern physicists are confident that the earth will crash into the sun in about one billion years (when the sun runs out of hydrogen and expands.) It is, thus, almost certain that human beings, or other species descending from us, will not be around *forever*, no matter how well we look after the planet. Moreover, if we were to escape from the planet before it crashes into the sun, we would also have to consider the fact that our galaxy will crash into the Andromeda galaxy in about four billion years. So the prospects for what Jonas in his formulation of the imperative of responsibility calls "indefinite continuation" look dim. It thus seems somewhat unclear why Jonas thinks we should assign *infinite* negative value to the extinction of humanity. If we think all humans have the same moral value, and the number of humans who will ever live is bound to be finite, and no single individual has infinite value, and the negative value of the whole is no less than the sum of its parts, then the negative value we ought to assign to the extinction of humanity is some unknown but large *finite* negative number. The upshot of all this is that Jonas's decision theoretic argument does not seem to go through.

However, it is worth stressing that Jonas's work tallies well with considerations emphasized by contemporary defenders of the precautionary principle (see chapter 10). It is, therefore, no surprise that the imperative of responsibility has been influential for the development of the precautionary principle as we know it today. So instead of reading Jonas literally, we could think of his argument as a general appeal to risk aversion. On such a reading, we should avoid taking unnecessary existential risks, including risks related to catastrophic environmental degradation and nuclear war.

CASE 14-2

A Techno-fix for Global Warming?

Engineers have suggested that the best way to tackle global warming is to develop "techno-fixes" that solve the problem. The term *geoengineering* refers to technologies designed to deliberately manipulate the earth's climate. Engineers are currently exploring the possibility of countering climate change with at least two forms of geoengineering: Solar Radiation Management (SRM) and Carbon Dioxide Removal (CDR).

SRM (see Figure 14.2) is a set of (future) technologies that aim to reflect some of the sun's energy back into space and thereby prevent the temperature on earth from increasing. There are various proposals for how to do this. According to researchers affiliated with the Oxford Geoengineering Programme, it might be feasible to introduce small reflective particles in the stratosphere. The particles could be dropped from specially designed airplanes flying at a very high altitude. Another option could be to use chemicals that increase the reflectiveness of clouds. A third option

could be to place large reflectors in space that block a small portion of sunlight before it reaches Earth. None of these technologies has actually been tried and tested, but a significant number of experts *believe* they would work.

The aim of CDR technologies is to actively remove carbon dioxide from the earth's atmosphere. This might be easier than blocking sunlight from reaching the atmosphere. A simple example of CDR would be to plant large numbers of trees, that is, use photosynthesis for transforming carbon dioxide and water to oxygen and carbohydrates (sugar). Another option could be to build large machines that capture carbon dioxide in the atmosphere and store it in the ground. Researchers in the Netherlands have proposed that empty fossil fuel reservoirs could be filled with carbon dioxide captured in this way. Yet another option could be to fertilize the ocean in a few select areas, which would draw down carbon dioxide from the atmosphere.

Figure 14.2
Solar Radiation Management is an example of a "techno-fix" for climate change that aims to reflect some of the sun's energy back into space and thereby prevent the temperature on earth from increasing.
Source: iStock by Getty Images.

Technological pessimists question the idea that geoengineering would be an appropriate response to global warming. They reject geoengineering *because* it is a techno-fix that could at best be a temporary solution to what is ultimately a man-made problem. It would, they believe, be naïve to think that we can fully control these technologies. The thought that geoengineering is the solution to global warming shows that we are suffering from hubris. The negative side effects of geoengineering may be enormous, and may not be worth taking.

Technological optimists respond that the worries raised by pessimists can be resolved by making sure the technology works as intended. We have to give the engineers the money and resources they need. For all we can tell now, geoengineering might be the best option we have for mitigating the effects of global warming. It is true that the cause of the problem is our own behavior, but it is unrealistic to think that we can turn back the clock and prevent global warming with social intervention. We have to be realistic about the planet's future; and in light of all the evidence available to us, we have no good reason to resist the promises of geoengineering.

Thinkers who take a critical attitude to technology point out that some forms of geoengineering seem to be more attractive than others. For instance, the risks with planting trees seem to be very limited; but if we fertilize the oceans or release reflective particles in the stratosphere, then the risk of unforeseen side effects is likely to be greater. Another advantage of planting trees is that this a reversible intervention. We can also stop using this technology whenever we want. However, the consequences of fertilizing the oceans or releasing reflective particles in the stratosphere are more difficult to predict. Although they are believed to be reversible, they may turn out not to be so. The upshot is that some methods of geoengineering seem to be more attractive than others. We have no reason to reject engineering outright, but we should also be cautious not to accept any unreasonable risks.

We also have to think critically about whether we would have to obtain informed consent from those affected by geoengineering. Would it be appropriate to hold a global election in which all members of the current generation have voting rights? If so, who should represent the interests of future generations? A more reasonable option could be to let the United Nations decide on matters related to geoengineering. This might, perhaps, be preferable to letting individual countries who want to try geoengineering do so without asking other nations for consent.

Discussion question: Some believe that a geoengineering and other "techno-fixes" of the climate problem can at best be a temporary solution to what is ultimately a man-made problem. In the long run, more fundamental changes to society and our lifestyle are needed for addressing this existential threat to humanity. Do you agree? Give arguments for and against.

BETWEEN PESSIMISM AND OPTIMISM

Hans Jonas's attitude to modern technology is, in a certain sense, extreme. His imperative of responsibility is motivated by a concern for the *negative* consequences of technology. As explained previously, he thinks that even if no catastrophic consequences were to actually materialize, the mere fact that they *could* materialize is sufficient reason for being pessimistic and cautious about modern technology.

Another extreme position is to adopt the happy-go-lucky attitude advocated by technological optimists. Technological optimists stress that, all things considered, humans are much better off today than ever before, which is largely due to technological innovations. The optimist admits that not every single technology has turned out to be beneficial, but modern technology *as a whole* has brought us enormous benefits.

Technological optimists believe that many problems we face in today's society can be solved by new technological innovations, or "techno-fixes." Consider that many elderly people need assistance with their daily activities. Rather than trying to change the social structure that caused this problem in the first place (social isolation of the elderly), technological optimists argue that modern care robots can solve the problem. By developing robots that assist the elderly with their household activities, the need for human care decreases.

Another example is online bullying and harassment on social media. Instead of addressing the underlying social causes, companies such as Gavagai have developed a software known as a "Troll Hunter," which they describe as "a plugin for automatically moderating messages, and blocking those that contain hateful, degrading, or slanderous, language."[8] This is yet again an example of a techno-fix of a social problem.

Critics of technological optimism do not deny that modern technology has enabled us to improve our living conditions, but they question the claim that we are much better off than our ancestors. They also question the idea that the negative side effects of modern technology can be solved with techno-fixes. In a society that seeks to address problems that are of social nature with robots and software, something valuable has been lost.

As noted at the beginning of the chapter, the *critical* approach to technology is an intermediate, less extreme position in this debate. Rather than making bold claims about the positive or negative value of technology as such, defenders of the critical approach argue that we should evaluate each and every new or existing technology individually. We can, of course, suggest design improvements or take other measures whenever that is motivated, but we should refrain from making general claims about the value of technology as such. A restaurant critic should praise or criticize individual restaurants (and sometimes individual dishes on the menu), but it would be pointless to praise or criticize all restaurants in the world collectively.

Engineers who take a critical approach to technology can apply several of the ideas discussed in this book for evaluating new and existing technologies. Consider, for instance, the list of questions mentioned in chapter 13, which Ibo van de Poel thinks we should ask about social experiments. The very same list could be used for formulating critical questions about individual technologies *without* making the controversial claim that technologies are social experiments. The following are some examples of critical questions we could ask about the care robot discussed previously:

1. Have the users of this technology given their informed consent? If not, is this a case in which no informed consent is needed?
2. Does this technology violate anyone's privacy? If so, is this violation of privacy necessary and well-motivated?
3. Are the benefits and costs of this technology distributed fairly in society? If not, can we redesign the technology such that the distribution becomes less unfair?

CASE 14-3

Amish Values

The Amish movement is a Protestant Christian church founded by Jakob Amman in Switzerland around 1680. The Amish fled religious persecution in Europe in the early eighteenth century and settled in Pennsylvania. Today about three hundred thousand people live in Amish communities spread over twenty-five states in the United States and Canada.

A distinctive feature of Amish culture is its aversion against modern technology. In Amish homes, there are no TVs, computers, telephones, or washing machines. Instead of electric lightbulbs, the Amish rely on gas lamps and natural light from windows. The Amish are widely known for not owning or driving cars. They prefer horse-drawn buggies (see Figure 14.3). However, the Amish do not believe there is anything wrong with modern technology *as such*. Technology is not considered to be an evil in itself. From a religious point of view, there are few differences between the Amish and other Protestants.

The Amish base their evaluation of new technologies on instrumental means-end reasoning, just like most of us; but their conclusion is different because their *values* (ends) differ from ours. For instance, if you (a non-Amish person) consider buying a new technological gadget, you will probably ask yourself, "How will this new gadget help me achieve my goals? Will it help me do things I care about?" The Amish ask the same question, but their conclusion is different because they do not strive for the same materialistic goals as other people. As an Amish minister interviewed by Wetmore (2007) put it, "We try to find out how new ideas, inventions, or trends will affect us as a people, as a community, as a church. If they affect us adversely, we are wary. Many things are not what they appear to be at first glance. It is not individual technologies that concern us, but the total chain."

The Amish value modesty, humility, equality, and social cohesion. They avoid almost everything

Figure 14.3
An Amish buggy in Lancaster County, Pennsylvania.
Source: Nicholas A. Tonelli.

(Continued)

that would make them stand out as individuals. No one is, for instance, permitted to dress in ways that make them look richer or more beautiful than others. The Amish prefer plain clothes that look the same as those their ancestors wore when they arrived in America. The point about social cohesion is particularly important, and is helpful for explaining how the Amish think about technology. Somewhat crudely put, technologies that help keep Amish families and communities together are permitted, while technologies that pose a threat to social cohesion are rejected. Electricity is, for instance, permitted when it is used for cooling locally produced dairy, because dairy that is not refrigerated cannot be sold to the "English" (this is what the Amish call all their English-speaking neighbors). Using electricity for refrigerating dairy can thus be seen as a pragmatic way of promoting Amish values, based on traditional means-ends reasoning.

A similar point can be made about how the Amish think about telephones. Telephones are banned in Amish homes because, as Wetmore's interviewee put it, "if everyone had telephones, they wouldn't trouble to walk down the road or get in the buggy to go visiting anymore." Note, again, that there is nothing wrong with telephones as such. It is rather the fear for instrumental effects on real-world social interactions that explain the ban. Interestingly, some Amish districts permit the use of public telephones located outside the home in special Amish phone booths, which are used for, for example, contacting English people or for carrying out business transactions with people far away, which is necessary for sustaining the Amish way of life. Similar considerations explain why the Amish prefer to use buggies over faster and more comfortable automobiles:

> Buggies are a social equalizer because they are uniform, free from excess bodywork and color, and because one buggy cannot be made significantly faster or slower than another. Automobiles, on the other hand are

criticized for providing an abnormal sensation of power that can be used to not only show up one's neighbors, but to abandon them altogether. (Wetmore 2007: 16)

For similar reasons, the Amish refuse to insure their homes. A family who knows that its house is uninsured is more likely to preserve stable social relations with their neighbors, so they can get help to rebuild their house if it is destroyed in a fire or windstorm. Not taking out insurance can thus be seen as a way of promoting Amish values.

The decision-making process in Amish communities is democratic. Every time a change to the "Ordnung" (the rules) that governs life in Amish communities is proposed, all members of the local district get together to discuss the proposal. Each district consists of about thirty to sixty families. All men and women have one vote each. A change in the Ordnung, such as permission to use a new technology, is rejected if two or more members oppose it. This conservative voting procedure explains why the Amish way of life has changed so little since the nineteenth century: it rarely happens that almost everyone agrees on a proposal to revise the Ordnung.

Needless to say, the Amish cannot be characterized as technological optimists, but are they technological pessimists, or advocates of the critical attitude? Although it is tempting to conclude that they are technological pessimists, it would probably be equally reasonable to argue that they defend an extreme version of the critical approach. They carry out separate evaluations of each and every new technology and almost always come to the conclusion that the new technology would not promote the values they care about; but they do not claim that *all* technologies, or technology *as such*, should be rejected.

Discussion question: Do you think the Amish are (on average) less happy than you? If not, would you be willing to live without technological gadgets?

4. Can the users of the technology stop using the technology whenever they wish? Or is it likely that users will become addicted to it?

5. What is the environmental effect of this technology? Is it acceptable?

6. Does this technology bring about any irreversible effects? Although it may not always be wrong to implement technologies with irreversible effects, it is worth thinking twice before we do so.

The answers to these questions may not fully determine the moral analysis of care robots, but they encourage us to think critically and make conscious decisions about whether we want to accept this new technology or not.

REVIEW QUESTIONS

1. Summarize the three attitudes to technology discussed in this chapter: pessimism, optimism, and the critical attitude.
2. What are the best arguments for and against these three views?
3. Formulate, by using your own words, Hans Jonas's imperative of responsibility.
4. Discuss similarities and differences between Jonas's imperative of responsibility and Kant's categorical imperative.
5. Discuss similarities and differences between Jonas's imperative of responsibility, the risk-benefit principle, and the precautionary principle.
6. Explain how defenders of the critical approach may wish to apply the principles proposed by van de Poel for evaluating social experiments. What is the role of these principles in the critical approach?
7. What is geoengineering, and what are the promises and risks of these technologies?
8. Explain how technological pessimists, optimists, and defenders of the critical approach think about geoengineering.

REFERENCES AND FURTHER READINGS

Boudette, N. E., and B. Vlasic. September 12, 2017. "Tesla Self-Driving System Faulted by Safety Agency in Crash." *New York Times.*

Jonas, H. 1985. *The Imperative of Responsibility: In Search of an Ethics for the Technological Age.* Chicago: University of Chicago Press.

National Transportation Safety Board. 2017. "Collision between a Car Operating with Automated Vehicle Control Systems and a Tractor-Semitrailer Truck near Williston, Florida May 7, 2016." Washington, DC: Accident Report NTSB/HAR-17/02 PB2017-102600.

Peterson, M. "The Value Alignment Problem: A Geometric Approach," forthcoming in *Ethics and Information Technology.*

Russell, S. 2016. "Should We Fear Supersmart Robots?" *Scientific American* 314, no. 6: 58–59.

Stilgoe, J. 2016. "Geoengineering as Collective Experimentation." *Science and Engineering Ethics* 22, no. 3: 851–869.

Wetmore, J. M. 2007. "Amish Technology: Reinforcing Values and Building Community." *IEEE Technology and Society Magazine* 26, no. 2: 10–21.

NOTES

1. See Taylor, M. (2016). "Self-driving Mercedes-Benzes will prioritize occupant safety over pedestrians", Retrieved January 26, 2018, from https://blog.caranddriver.com/self-driving-mercedes-will-prioritize-occupant-safety-over-pedestrians.

2. References for all quotes in this case description can be found in M. Peterson "The Value Alignment Problem: A Geometric Approach," which is the main source for Case 14-1.

3. IEEE, December 2017, version 2.

4. Hans Jonas, *The Imperative of Responsibility: In Search of an Ethics for the Technological Age* (Chicago: University of Chicago Press, 1985), ix–x.

5. Ibid. 11.

6. Ibid.

7. Jonas, *The Imperative of Responsibility*, 37–38, italics in original.

8. Gavagai.se (last accessed June 25, 2017).

CHAPTER 15

The Ethics of Artifacts

Engineers design, build, and interact with a myriad of artifacts. Cell phones, airplanes, ultrasound scanners, nuclear weapons, and self-driving cars are very different entities, but they are all examples of physical objects created by engineers. In the academic literature, the term *technological artifact* refers to any object intentionally created by human beings.

What is the role of technological artifacts in ethics? Are they morally neutral and inert objects that play no significant role in moral reasoning whatsoever, or do technological artifacts have moral properties of their own? Consider, for instance, Lee Harvey Oswald's assassination of President Kennedy on November 22, 1963. While it is true that Lee Harvey Oswald would not have been able to shoot President Kennedy without a rifle, it seems fairly obvious that the rifle itself did not deserve any blame for Kennedy's death. The fact that Oswald was in possession of a rifle enabled him to do something he wouldn't have been able to do otherwise, but this does not support the conclusion that *the rifle* had any moral properties. Let us call this the *common-sense view* about artifacts. On this view, the rifle was not morally complicit in the killing of Kennedy in any way or capacity. Irrespective of whether Oswald was solely responsible for the assassination, or was assisted by someone else, or was completely innocent, the moral agent who killed Kennedy was not a technical artifact. Kennedy was shot by a human being.

Somewhat surprisingly, several philosophers of technology reject the common-sense view. These philosophers believe in (slightly different versions of) what we may call the *strong view*. Defenders of the strong view dismiss the idea that technological artifacts are morally neutral or inert objects. They believe that technological artifacts play a much more prominent role in ethics. In this chapter, we shall take a closer look at a couple of different versions of the strong view.

WHAT'S SO SPECIAL ABOUT TECHNOLOGICAL ARTIFACTS?

To understand why some thinkers feel attracted by the strong view and the idea that "the things themselves" play an important role in ethics, it is helpful to consider ways in which technological artifacts *differ* from other objects. To make the comparison concrete, we can compare the rifle Oswald used for shooting Kennedy with innate natural objects such as rocks and mountains.

The Designer's Intention

Rifles and other technological artifacts are always created intentionally by a designer for a specific purpose. Unlike natural artifacts, technological artifacts are never the product of random processes. Oswald's rifle was designed with the aim of enabling its user to kill people. Rocks, stones, and rivers are not intentionally created by anyone for any purpose at all. It is of course true that it's sometimes possible to kill someone by throwing a stone at them, but this just shows that the stone can be used as a piece of technology.

Needless to say, the fact that technological artifacts are intentionally created by a designer for a specific purpose does not show that the common-sense view is false. To show this, we would have to demonstrate that the designer's intentions are "contagious," in the sense that they are somehow transmitted to the technological artifact being designed.

CASE 15-1

Topf & Söhne and the Holocaust

During World War II, the Nazis systematically murdered some six million Jews in concentration camps in, for example, Auschwitz, Treblinka, Sobibor, Chelmno, and Belzec. Although engineers were not *directly* responsible for this atrocity, their knowledge and skills were crucial for enabling the Nazis to murder so many people in such a short period of time. Toward the end of the war, most victims were killed with gas (Cyclone B) and cremated. Topf & Söhne, a family-owned company managed by Ludwig and Ernst-Wolfgang Topf, was the largest supplier of crematorium furnaces to the concentration camps.

After the war, Ludwig and Ernst-Wolfgang Topf claimed that they had had no knowledge of what their furnaces (see Figure 15.1) had been used for, although they admitted they knew some furnaces had been delivered to concentration camps. Modern historians are convinced that Ludwig and Ernst-Wolfgang lied. Senior engineer Kurt Prüfer and other employees admitted (see the following)

that they knew perfectly well what the Nazis used the crematorium furnaces for. Ludwig Topf committed suicide on May 31, 1945, but Ernst-Wolfgang lived a relatively peaceful life in freedom in West Germany until his death in 1979. He was never held accountable for having manufactured and delivered the furnaces in which millions of people killed by the Nazis were cremated. In the 1950s, Ernst-Wolfgang started a new company with the same name and product line. It went bankrupt in 1963 as photos of dead corpses next to Topf-made furnaces in concentration camps appeared in German newspapers.

The furnaces delivered to the Nazis were designed by Topf & Söhne's senior engineer Kurt Prüfer. He was responsible for, among other things, the innovative and technically advanced "triple-muffle" furnace designed in 1940. Prüfer had studied structural engineering for six semesters in Erfurt before joining Topf & Söhne. After the war, he initially insisted that he had not known what the ovens were

Figure 15.1
Crematorium furnaces used by the Nazis in Buchenwald.
Source: Ad Meskens.

being used for; but when interrogated by Soviet officials in 1946, he changed his testimony and admitted that he had visited Auschwitz three times in 1943 and that he knew what was going on there:

> I have known since spring 1943 that innocent human beings were liquidated in Auschwitz gas chambers and that these corpses were subsequently incinerated in the crematoriums.[1]

The actions of Prüfer and other engineers at Topf & Söhne were *paradigmatically* wrong. Engineers should not design and build devices that they know will be used for cremating innocent civilians murdered by a regime with no respect for basic human rights. It is paramount that every new generation of engineers is aware of the role engineers played for enabling the Nazis to murder millions of innocent people. Without the help of engineers, it would not have been feasible for the Nazis to do what they did.

However, this tragic and painful example also highlights the importance of the designer's intention. Prüfer designed crematorium ovens with the *intention* of helping the Nazis fulfill their desire to kill and cremate as many innocent human beings as possible.

Discussion questions: Was the evil intention with which Prüfer designed his furnaces "transmitted" to the technological artifact itself? That is, were the crematorium furnaces somehow "loaded" with negative moral value? Or were the furnaces morally neutral objects with no moral properties of their own? What is in your opinion the best objection to the claim that Prüfer's evil intention was "transmitted" to the technological artifacts he designed, that is, to his triple-muffle crematorium furnace?

Unforeseen Use and Side Effects

Many technological artifacts are used in ways that differ radically from the way intended by their designers. As mentioned in previous chapters, the telephone was originally designed by Bell in 1876 as an aid for people hard of hearing, but that is not how it is used today. Another example is the cell phone. It was originally designed for making phone calls, but most of us use it primarily for sending text messages and browsing the Internet. Another example is the revolving doors we find in hotels and other large buildings. Revolving doors were originally designed in the nineteenth century to keep cold air out during the winter months, but an unforeseen side effect was that it became more difficult for disabled citizens to enter the building. No one could foresee this effect when the technology was first introduced.

The fact that there tends to be a gap between the way technological artifacts are *designed* to be used and the way they are *actually* used is sometimes taken to indicate that, from a moral point of view, artifacts have a life of their own (in a metaphorical sense). Because the designer cannot reasonably foresee how an artifact will be used, we cannot hold the designer accountable for how it is in fact used.

The Mediating Role of Technological Artifacts

Technological artifacts *mediate* our actions as well as our perception of the world. The word "mediate" literally means "placing in the middle." For an example of how technologies mediate our perception of the world, imagine yourself wearing a pair of blue sunglasses. The lenses are "in the middle" between you and the external world; but because they are blue, you see the world in slightly different shades of blue. Another example is a microscope used by a scientist for studying tiny objects such as molecules and cells. The microscope mediates the scientist's perception of the world by enabling him or her to see things that are otherwise impossible to see.

For an example of how actions are mediated by technological artifacts, consider Oswald's rifle. This technological artifact enabled Oswald to kill Kennedy. He wouldn't have been able to do so without it. The rifle thus mediated Oswald's actions in the sense that it created new options for him. Another example is the Internet. Because the Internet makes it easier for people to express vulgar and uncivilized opinions without being held accountable, many Internet users do so. The Internet changes the way we communicate with each other; this may not always be for the better, but it is hard to deny that the Internet has changed many of our activities quite dramatically.

CAN ARTIFACTS BE VALUE-LADEN?

What moral conclusion, if any, follows from the observation that technological artifacts are designed with a specific intention in mind, used in ways the designer could not have foreseen, and mediate our actions and perception of the world?

Philosopher Langdon Winner focuses on the first feature of technological artifacts mentioned previously, that is, the fact that they are intentionally designed for a specific purpose. According to Winner, the fact that technological artifacts are intentionally designed objects supports the conclusion that the artifacts themselves literally *embody* the designer's moral and political values.

To explain what leads him to this conclusion, it is helpful to consider the bridges (see Figure 15.2), buildings, and parks in New York designed by the famous city planner Robert Moses (1888–1981). According to Winner, Moses was a racist. To claim that someone is a racist can be controversial; but in what follows, we will assume for the sake of the argument that Winner is right about this. (It is worth keeping in mind that some historians believe Moses was not a racist.[2]) Here is Winner's analysis of Moses's work:

> Anyone who has traveled the highways of America and has become used to the normal height of overpasses may well find something a little odd about some of the bridges over the parkways on Long Island, New York. Many of the overpasses are extremely low. [. . .] Robert Moses, the master builder of roads, parks, bridges, and other public works from the 1920s to the 1970s in New York, had these overpasses built to specifications that would discourage the presence of buses on his parkway. [. . .] The reasons reflect Moses's social-class bias and racial prejudice. Automobile-owning whites of "upper" and "comfortable middle" classes, as he called them, would be free to use the parkways for recreation and commuting. Poor people and blacks, who normally used public transit, were kept off the roads because tall buses could not get through the overpasses. [. . .] Many of his monumental structures of concrete and steel *embody a systematic social inequality, a way of engineering relationships among people that, after a time, becomes just another part of the landscape.*[3]

Winner's point is that Moses's intention to segregate people by designing unusually low overpasses "infected" the artifacts with his own values. On Winner's view, there is systematic social inequality and racial prejudice built into the overpasses because those values guided Moses's design of the overpasses. For another example, consider

Figure 15.2
Robert Moses deliberately designed low overpasses that would reduce the number of buses on his parkway. According to Langdon Winner, these low overpasses "embody a systematic social inequality."   Source: iStock by Getty Images.

the apartheid system in South Africa, which ended with the release of Nelson Mandela in 1990. In the years before Mandela was released, white and nonwhite people lived in parallel worlds. The apartheid regime built separate schools, healthcare units, and restaurants for white and nonwhite people. Winner would say that the social values of the apartheid system were *engineered* into nearly all buildings in the South African society.

Should we accept this analysis of how moral and political values are, quite literally, built into steel and concrete? Unsurprisingly, Winner's proposal is controversial.

To start with, it could be objected that social structures may *change* at any point in time, but this does typically not lead to any change of the steel and concrete designed to reflect the designer's values. If Winner is right, the low overpasses designed by Moses would still embody social inequality and racial prejudice, even though almost everyone can afford a car in the United States today. This implication of his analysis is, perhaps, a bit strange. Why is the social inequality and racial prejudice still present in the low overpasses if they no longer create any discrimination? It seems odd to maintain that the overpasses still embody the original, racist values in a world in which everyone can afford to travel by car.

Another problem is that it is difficult to make sense of the idea that values can be embodied in a physical object. What does it *mean* to say that concrete and steel embody moral or political values? If you examine the object, it will surely look like any other physical object. The values are nowhere to be seen unless you know the designer's intentions. However, if we cannot "see" the values embodied in an artifact without knowing the designer's intention, it might be more appropriate to say that the values are "in the eye of the beholder" rather than embodied in the object.

CAN ARTIFACTS BE HYBRID AGENTS?

The observation that technological artifacts mediate our actions has led French sociologist Bruno Latour (1947–) to believe that artifacts, together with human beings, form a special kind of hybrid agents, which he calls "actants." Here is how Latour formulates his view:

> Here is an example of what I have in mind: the speed bump [see Figure 15.3] that forces drivers to slow down on campus, which in French is called a "sleeping policeman." . . . The driver modifies his behavior through the mediation of the speed bump: he falls back from morality to force. . . . On the campus road there now resides a new actant that slows down cars.[4]

Latour does not claim that the *speed bump itself* is a moral agent or "actant." It is not the speed bump that performs the action of slowing down cars. Latour's view is, rather, that it is the *combination* of the speed bump and the people who design and use it that slows down cars. As noted, we can think of this combination of a technological artifact and a human being as a hybrid agent.

If we apply this line of reasoning to the assassination of President Kennedy, we could say that it was Oswald-and-his-rifle who shot Kennedy. The rifle itself did not kill anyone, nor did Oswald, but the hybrid agent consisting of Oswald-and-his-rifle did.

Figure 15.3
According to Bruno Latour, a speed bump is an "actant" that forces drivers to slow down; the speed bump is no passive or morally neutral entity. Source: iStock by Getty Images.

Latour's theoretical framework is called the Actor-Network Theory. One of its key claims is that the boundary between *objects* (such as guns and speed bumps) and *subjects* (a murderer or a bus driver on campus) is blurry. On Latour's view, we should reject the common-sense view according to which objects and subjects are radically different entities clearly separated from each other. According to Latour's Actor-Network Theory, Oswald-and-the rifle is, in a literal sense, one and the same "thing." It was the rifle together with Oswald and all the other relevant components of the "network" that killed Kennedy.

Is Latour's Actor-Network Theory convincing? The best way to assess his radical view is, arguably, to perform an intellectual cost-benefit analysis. What are the intellectual gains of accepting Latour's view, and what would it cost us intellectually? Would our new set of beliefs be more coherent than our original beliefs?

The upside of Latour's theory is that he can explain, in a quite straightforward manner, why technological artifacts matter in ethics. It seems clear that numerous ethical problems triggered by new technologies require our attention. Latour is able to explain what role the technological artifacts themselves play in these ethical inquiries, whereas defenders of the common-sense view must at some point conclude that technological artifacts are, strictly speaking, not at all that special from a moral point of view.

The downside of Latour's theory is that it is very radical. It forces us to revise some of our most fundamental assumptions about ethics and moral agency; but it is often wise to be conservative and revise one's most fundamental beliefs only if that is absolutely necessary for accommodating some new, convincing piece of evidence. This is a general methodological principle that applies to all academic fields. We know that our current beliefs about the world are fairly coherent, so if we make radical revisions, there is a significant risk that we end up with fundamentally incoherent beliefs. At present, it seems that the totality of all our beliefs would not become more coherent if we were to incorporate Latour's radical proposal.

Another problem with Latour's view is that it is difficult to understand what it means, exactly, that Oswald-and-his-rifle killed Kennedy. The rifle has no consciousness or other properties we normally associate with moral deliberation. When we analyze the hybrid entity consisting of Oswald-and-his-rifle, it seems plausible to say that "all the morality" is somehow located in Oswald, rather than the rifle.

CASE 15-2

The Great Firewall in China and Technological Mediation

The Great Firewall Project is an Internet censorship and surveillance project in China controlled by the ruling Communist Party. By using methods such as IP blocking, URL filtering, DNS filtering, and redirection, the Chinese Ministry of Public Security is able to block and filter access to information deemed to be politically dissident or "inappropriate" for other reasons. As part of this policy, searching with all Google search engines was banned in Mainland China on March 30, 2010. The project started in 1998 and is still in operation.

In a sample of 564 students taking a course in engineering ethics at Texas A&M University, 81 percent reported that in their opinion, the Great Firewall raises concerns about the autonomy of Internet users in China.[5]

Latour would argue that the Great Firewall is an example of *technological mediation*. By censoring the Internet, the Chinese authorities influence the actions and perceptions of people in China. Citizens who are unable to access information about, for instance, the massacre in Tiananmen Square in 1989 are less likely to complain about violations of human rights. This changes people's perception of the world and makes it easier for the Communist Party to suppress dissident opinions.

Some forms of technological mediation are morally unproblematic, whereas others are not. If we believe that the Great Firewall is a morally problematic form of technological mediation, it does not follow that we should not focus our concerns exclusively on the Chinese authorities. According to Latour's Actor-Network Theory, it is the Chinese authorities *in combination with* the technology used for filtering the Internet that is responsible for this morally problematic form of technological mediation. We can compare this with Latour's speed bump example: if we accept Latour's claim that "on the campus road there now resides a new actant that slows down cars,"[6] it is also reasonable to accept the claim that the Great Firewall is a new actant that blocks and filters access to information deemed to be politically dissident. A possible objection could be that the responsibility of the individual human beings who took the political decision to censor the Internet runs a risk of being diminished by Actor-Network Theory.

Discussion question: Foreigners visiting China can create a Virtual Private Network tunnel that enables them to send sensitive or illegal information through the Great Firewall. Would it be morally right to do so? If so, who is the moral agent responsible for this act? The user, the designer of the technology, the VPN technology itself, or some combination thereof?

TECHNOLOGY AS A WAY OF REVEALING

Martin Heidegger (1889–1976) is probably the most famous philosopher of technology of all times. He was rector of the University of Freiburg from 1933 to 1934. He was also a member of the Nazi party until 1945. A series of recently discovered letters shows that he was deeply devoted to the Nazi ideology. However, the fact that he was a Nazi does not entail that his nonpolitical views must have been false, nor does the fact that he is famous prove his philosophical views to be true. We should accept Heidegger's philosophical views if and only if we have a good epistemic reason to believe they are correct.

Heidegger's philosophical positions can be interpreted in different ways. His style is notoriously difficult to understand and some of his key terms have multiple meanings. In "The Question Concerning Technology," Heidegger writes that

> Technology is . . . no mere means. Technology is a way of revealing . . . an airliner that stands on the runway is surely an object. Certainly. We can represent the machine so. But then it conceals itself as to what and how it is. *Revealed, it stands on the taxi strip only as standing-reserve, inasmuch as it is ordered to insure the possibility of transportation.*[7]

One (of several possible) interpretations of this passage is that Heidegger believes the morally relevant properties of an airliner are grounded in the fact that it gives us "the possibility of transportation." Let us explore this interpretation a bit further, regardless of what other interpretations might be possible. According to this straightforward interpretation, technological artifacts are morally relevant in the sense that they give us access to new options for action that were unavailable to us before the technology became available. For instance, the development of modern airliners made it possible to travel long distances quickly which was not possible at the beginning of the twentieth century. Similarly, the development of nuclear weapons made it possible to kill more people than ever; and the rifle Oswald used enabled him to assassinate Kennedy from a great distance; and so on.

Our interpretation of Heidegger does not entail that a technological artifact is (part of) any moral agent. The artifact itself performs no actions, nor does the artifact together with any human being do so; humans perform all actions. However, the availability of the artifact "reveals" a "standing-reserve" for us to do things we could not otherwise have done. Artifacts are thus morally relevant in so far as they create new, morally relevant options for action.

Two contemporary philosophers of technology, Christian Illies and Anthonie Meijers, explore a view similar to Heidegger's. They argue that, everything else being equal, it is always better to have more options for action rather than fewer. Many modern technologies make it possible for us to perform good actions that were previously unviable, such as treating cancer with radiation therapy or driving electric cars. Unlike Heidegger, their view can thus be described as an example of technological optimism rather than pessimism. However, according to Illies and Meijers, the fact that it is nowadays possible to do things that were previously impossible also places some new responsibilities on engineers. Illies and Meijers distinguish between what they call the engineer's *first-* and *second-order* responsibilities. Everyone (not just engineers) has a first-order responsibility to choose a morally right alternative from the set of alternatives available at each point in time. Oswald did not fulfill his first-order responsibility because murdering the president was not a morally permissible option. Oswald should have refrained from pulling the trigger and expressed his dissatisfaction with the president in some less violent way.

However, in addition to their first-order responsibilities, engineers have a special second-order responsibility to create new, morally good actions for ordinary users of technology. Here is an example: an engineer who develops a safer and more environmentally friendly car creates a new alternative for car buyers. This enables ordinary consumers, if they so choose, to perform actions that are morally better than the ones previously available to them. If the engineer had instead developed a car that is unsafe and harms the environment, he or she would not have fulfilled the second-order responsibility to create new morally good options for action.

The notion of second-order responsibility explains what it is that makes engineering special from a moral point of view: the engineer indirectly decides what options become available to others. If engineers develop technologies that present the rest of humanity with bad alternatives ("cars that pollute 10 percent more than today"), then this is a moral disaster that affects all of us. In that sense, it is important that engineers fulfill their second-order responsibility to create morally attractive alternatives ("carbon-neutral cars") that ordinary users of technology can chose.

The view of technology sketched here comes with several problems, just like almost every philosophical position. First, one could object that second-order responsibilities do not seem to be at odds with the common-sense view. We can coherently claim that engineers have a moral responsibility to develop carbon-neutral cars *and* that cars themselves have no moral properties. The fact that new technologies make it possible to do things that were previously impossible may not entail that the artifacts themselves are good or bad from a moral point of view. We could explain whatever needs to be explained by highlighting the actions performed by human agents: it is morally praiseworthy of an engineer to develop a carbon-neutral car and it is also praiseworthy to buy and drive one; but the car itself has no moral properties. We can compare this with a mountain guide who has a responsibility to help his customers to find a safe route up the mountain. The mountain guide's actions can be right or wrong, as can the actions of the tourists; but the mountain itself arguably has no moral properties, even though its shape and other physical properties determine what options will be available to the climbers.

Another worry is that it is not easy to see why actions that are not *actually performed* would matter from a moral point of view. If a new action becomes available, then it seems that this should make a difference only if the new alternative is actually performed. However, the notion of second-order responsibilities assigns moral value to options that are never performed. The hydrogen bomb is a good example. No such bomb has ever been used, but should the mere existence of this weapon make us say that it changed the world for the worse? If so, why? Note that defenders of the common-sense view could easily say that it would be morally *wrong* to use the hydrogen bomb. They could also say that it would be wrong to *threaten* to use the bomb, and that it is wrong to use it as a tool for increasing the nation's bargaining power. None of these claims require that we ascribe any moral properties to the mere existence of the bomb. So in what sense is the "standing-reserve" presented by the hydrogen bomb morally relevant?

REVIEW QUESTIONS

1. Are technological artifacts morally neutral objects that play no significant role in moral reasoning whatsoever? If not, in what sense are they relevant?
2. Who shot President Kennedy in Dallas: was it Lee Harvey Oswald, or his rifle, or Oswald *and* his rifle? (Make sure that you fully understand this question before you attempt to answer it.)
3. Discuss the role of the designer's intention when evaluating technological artifacts.
4. What is "technological mediation" and what is the moral importance of this concept?
5. Did Robert Moses's intention to segregate people on Long Island by designing unusually low overpasses "infect" those artifacts with his values? Why or why not?
6. Bruno Latour believes that "the speed bump that forces drivers to slow down on campus [is] a new actant that slows down cars." What does he mean by this? What is an "actant"?
7. Martin Heidegger argues that "an airliner . . . stands on the taxi strip only as standing-reserve, inasmuch as it is ordered to insure the possibility of transportation." What does he mean by this? What is a "standing-reserve"?
8. Is the fact that Martin Heidegger was a member of the Nazi party until 1945 relevant for evaluating his contributions to the philosophy of technology?

REFERENCES AND FURTHER READINGS

Bernward, J. 1999. "Do Politics Have Artefacts?" *Social Studies of Science* 29, no. 3: 411–431.

Habermas, J. 1970. "Technology and Science as Ideology." In *Toward a Rational Society: Student Protest, Science, and Politics*. Chapter 6. Boston: Beacon Press.

Heidegger, M. 2009. "The Question concerning Technology." In *Technology and Values: Essential Readings*, edited by C. Hanks, 99–113. Malden, MA: John Wiley & Sons. First published 1954.

Illies, C. and A. Meijers. 2009. "Artifacts without Agency." *Monist* 92, no. 3: 420–440.

Katz, E. 2011. "The Nazi Engineers: Reflections on Technological Ethics in Hell." *Science and Engineering Ethics* 17, no. 3: 571–582.

McGinn, R. 2018. *The Ethical Engineer: contemporary Concepts and Cases*. Princeton, NJ: Princeton University Press.

Latour, B. 2009. "A Collective of Humans and Nonhumans: Following Daedalus's Labyrinth." In *Readings in the Philosophy of Technology*, by D. M. Kaplan. London: Rowman and Littlefield.

Verbeek, P. P. 2011. *Moralizing Technology: Understanding and Designing the Morality of Things*. Chicago: University of Chicago Press.

Wetmore, J. M. 2007. "Amish Technology: Reinforcing Values and Building Community." *IEEE Technology and Society Magazine* 26, no. 2: 10–21.

Winner, L. 1980. "Do Artifacts Have Politics?" *Daedalus* 109, no. 1: 121–136.

NOTES

1. R. McGinn, *The Ethical Engineer: Contemporary Concepts and Cases* (Princeton, NJ: Princeton University Press, 2018), 167.
2. Bernward Joerges provides a nuanced and detailed discussion of Moses's motives. He writes that Moses "never pursued explicitly racist schemes," although he admits that Moses was "what one might call a 'structural racist'." See p. 418 in Bernward Joerges, "Do Politics Have Artefacts?" *Social Studies of Science* 29, no. 3 (1999): 411–431.

3. Langdon Winner, "Do Artifacts Have Politics?," *Daedalus* 109, no. 1 (1980): 123–124, my italics.

4. Bruno Latour, "A Collective of Humans and Nonhumans: Following Daedalus's Labyrinth," in *Readings in the Philosophy of Technology*, ed. D. M. Kaplan (London: Rowman and Littlefield, 2009), 186–188.

5. M. Peterson, *The Ethics of Technology: A Geometric Analysis of Five Moral Principles* (New York: Oxford University Press, 2017), 158.

6. Bruno Latour, "A Collective of Humans and Nonhumans: Following Daedalus's Labyrinth," in *Readings in the Philosophy of Technology*, ed. D. M. Kaplan (London: Rowman and Littlefield, 2009), p. 188.

7. M. Heidegger, "The Question concerning Technology" in *Technology and Values: Essential Readings*, ed. C. Hanks, 99–113 (1954; repr., Malden, MA: John Wiley & Sons, 2009), 294–295, my italics.

CHAPTER 16

Sustainability

Engineering projects sometimes harm the environment. To what extent, if any, are engineers responsible for the environmental consequences caused by technologies they design, manufacture, or operate?

Ethical discussions of environmental issues are often couched in terms of "sustainability" and "sustainable development." In the Oxford English Dictionary, *sustainability* is defined as "The property of being environmentally sustainable; the degree to which a process or enterprise is able to be maintained or continued while avoiding the long-term depletion of natural resources."[1] In this chapter, we will investigate some ethical aspects of sustainability. We will give particular attention to what engineers can and should do if their attempts to find sustainable solutions to technological problems clash with other more mundane values, such as short-term economic profit.

THREE NOTIONS OF SUSTAINABILITY

The *Deepwater Horizon* case discussed on the next page (Case 16-1) raises several ethical issues concerning the lack of adequate training and oversight of the crew working on the platform. In this chapter, we will, however, ignore the organizational problems and focus exclusively on the ethical issues related to sustainability.

As noted previously, sustainability is sometimes defined in terms of the degree to which something can be maintained or continued without significant long-term depletion of natural resources. This definition is sometimes contrasted with other, more extensive definitions. According to the US Environmental Protection Agency (EPA), the depletion of social and economic resources also matters in discussions of

CASE 16-1

The *Deepwater Horizon* Oil Spill

Deepwater Horizon was a semi-submersible off-shore drilling platform built in South Korea in 2001. The platform was about 400 ft. long and 250 ft. wide and capable of drilling in waters up to 8000 ft deep. In the spring of 2010, *Deepwater Horizon* was operated under a lease contract by British Petroleum (BP) to conduct exploratory drilling at the Macondo Prospect in the Gulf of Mexico about forty miles off the coast of Louisiana.

On April 20, the crew of *Deepwater Horizon* was drilling at a water depth of approximately 5,000 ft. (1,500 m). The pressure in the well was very high. At 9:45 p.m., a lethal combination of methane gas, mud, and sea water erupted from the well. The gas

ignited, which caused a firestorm that was visible from more than forty miles (see Figure 16.1).

The official investigation concluded that one of the causes of the accident was the lack of so-called *centralizers* in the wellbore. A centralizer is a device that keeps the drill string at the center of the wellbore. The software used by the drilling operators predicted that twenty-one centralizers would be needed for controlling the position of the drill string. However, only six were available on the platform. The crew had ordered fifteen additional centralizers, which were delivered by helicopter. Unfortunately, the new centralizers were believed to be of the wrong size. John Guide, the

Figure 16.1
The *Deepwater Horizon* accident in 2010 killed 11 of the 126 crew onboard the platform. It is estimated that 4.9 million barrels of crude oil were discharged into the Gulf of Mexico.
Source: US Coast Guard.

well team leader, decided to go ahead with only six centralizers, to save time and money. The drilling engineer in charge of the operations, Brett Cocales, supported Guide's decision and argued that Guide was right about the "risk-reward equation." In his view, the risk incurred by using too few centralizers was worth taking given how much money it would have cost BP to delay the drilling operations.

The explosion killed 11 of the 126 crew onboard. The survivors tried to activate the so-called blowout preventer. When they realized that this would not work, a final attempt was made to prevent what would become an environmental disaster by activating a device known as the "blind shear ram." This also failed. The *Deepwater Horizon* sank thirty-six hours later, but it took almost three months to plug the well. During those three months 4.9 million barrels of crude oil (210 million US gal. or 780,000 m³) were discharged into the sea. This, of course, had a negative impact on the marine environment in the Gulf of Mexico. The *Deepwater Horizon* oil spill is the largest marine oil spill in history.

BP and the owner of the platform, a company called Transocean, agreed to pay $4 billion damages for violations of the US Clean Water Act. The 2016 Hollywood film *Deepwater Horizon*, starring Mark Wahlberg, is a dramatization of the tragic events at the drilling platform.

Discussion question: What is, in your opinion, the most important *moral* issue illustrated by the *Deepwater Horizon* disaster?

---

sustainability. Until the shift of power in Washington, DC in 2017, the EPA endorsed the following broad notion of sustainability:

> Sustainability creates and maintains the conditions under which humans and nature can exist in productive harmony, that permit fulfilling the social, economic and other requirements of present and future generations.[3]

It is not clear how we should understand the term "productive harmony." Can a large wind farm exist in "productive harmony" with nature? If it generates a lot of electricity, it is clearly "productive," but does it produce all this energy in "harmony" with nature? Anyone who has seen the wind farms in Texas or Kansas (which tend to dominate the landscape in some areas) might be inclined to answer this question in the negative, even though wind farms are often seen as paradigmatic examples of sustainable energy solutions.

Leaving this worry aside, the broad notion of sustainability developed by the EPA can be further divided into *weak* and *strong* notions of sustainability. (According to weak accounts, trade-offs between economic, social, and environmental consequences are permitted.) If, for instance, the *Deepwater Horizon* oil spill caused x units of damage to the natural environment in the Gulf of Mexico, this could be compensated for by an economic gain of y dollars, where x and y are some appropriate numbers. The overall sustainability of a world that is x units worse from an environmental perspective would be preserved if it came with an economic gain of y dollars. So, on this view, environmental degradation need not by itself lead to unsustainability. What matters is whether the depletion of natural resources exceeds the social and economic benefits derived from it.

According to defenders of strong sustainability, a loss of a few units in one of the three dimensions—environmental, social, and economic sustainability—cannot be compensated by gains to the other dimensions. Every loss is a loss no matter what benefits it brings about. This leaves us with three alternative notions of sustainability:

1. *Narrow sustainability*: Sustainability be achieved only by making sure that there is no significant long-term depletion of natural resources.
2. *Broad and weak sustainability*: Significant long-term depletion of natural, social, or economic resources are permitted as long as the total aggregated value is preserved; losses in one dimension can be compensated by gains in other dimensions.
3. *Broad and strong sustainability*: This can be achieved only by making sure that there is no significant long-term depletion of natural, economic, or social resources; losses in one dimension *cannot* be compensated by gains in other dimensions.

The difference between the second and third notions can be illustrated in a thought experiment known as *Global Manhattan* (see Figure 16.2).[4] Imagine a future world in which the entire planet is covered with highways, skyscrapers, airports, factories, and a few artificial parks for recreation. In such a world, there is no space left on the entire planet for natural forests, mountains, rivers, and the organisms we usually find in nature. If the economic and social resources generated by the factories, airports, and highways in Global Manhattan are large enough, then advocates of weak sustainability would have to concede that that world is *more* sustainable than the actual world. Defenders of strong sustainability would deny this and argue that the actual world is more sustainable, and perhaps also better. (Needless to say, "more sustainable" does not mean better, all things considered.)

Figure 16.2
Global Manhattan is a future world in which the entire planet is covered with highways, skyscrapers, airports, factories, and a few artificial parks for recreation. Would you like to live in a such a world?   Source: iStock by Getty Images.

CASE 16-2

Is Fracking a Sustainable Technology?

Between 2008 and 2016, the production of oil and gas in the United States increased by about 90 percent. In 2017, for the first time ever, the United States exported more natural gas than it imported. This huge increase in production was to a large extent made possible by the introduction of a new technology: hydraulic fracturing, or "fracking."

In some locations, oil and gas can be extracted by drilling a hole in the ground and pumping up the content from large underground reservoirs. Such underground reservoirs are, however, relatively rare. Fracking is a method for extracting hydrocarbons from bore wells in which no or little oil or gas can be extracted with traditional methods. A fluid (mostly water) is pumped into the bore well under high pressure. This makes the ground more porous, which in turn makes oil and gas flow to the bore well.

Fracking has been criticized for inducing man-made earthquakes. Only two earthquakes stronger than magnitude 3 struck Texas in 2008. However, in 2016, after fracking had boosted the state's production of hydrocarbons, Texas was hit by twelve earthquakes of the same magnitude. Other states have experiences similar man-made earthquakes.

Representatives of the oil and gas industry have been quick to point out that the fracking process itself is not the cause of most of the earthquakes. About 98 percent of the earthquakes actually occur *after* oil and gas has been extracted and waste water is pumped back into the bore well. A simple method for reducing the number of earthquakes is, therefore, to dispose of the waste water in some other way. Companies that have chosen to do so have not had any problems with induced earthquakes. However, because the waste water is radioactive, it is not always possible to find environmentally responsible

ways to dispose of it. Scientists have found concentrations of barium, strontium, and radium in waste water that are 200 times higher than normal. If the waste water were to be desalinated, then the level of radioactivity would become so high that the material would be classified as radioactive waste.

The upshot is that frackers are faced with two environmentally problematic options. They can either induce earthquakes by pumping waste water back into old bore wells or look for other ways of disposing radioactive waste water above ground. It is also worth keeping in mind that these are not the only environmental issues with fracking. The technology is noisy, pollutes the air, and can affect ground water levels—but fracking has also been a welcome boost to the economy.

In what sense, if any, is fracking a sustainable technology? Needless to say, all methods for extracting (and combusting) hydrocarbons are unsustainable if we apply the *narrow* notion of sustainability. Fracking diminishes the world's resources of hydrocarbons, full stop. However, if we prefer a broad and weak notion of sustainability, then fracking may actually be said to be a sustainable technology. The economic benefits have so far been, and will probably continue to be, significant. According to this line of reasoning, the positive economic effects of fracking outweigh its negative environmental impacts. However, fracking would not count as sustainable if we were to adopt the broad and strong notion because that notion does not allow for trade-offs between environmental and economic effects.

Discussion question: Whether fracking is a sustainable technology depends on what notion of sustainability we accept, as noted earlier. So what notion of sustainability is, in your opinion, the most plausible one, and why?

THE INSTRUMENTAL VALUE OF SUSTAINABILITY

The three notions of sustainability outlined in the preceding section stress the moral importance of not depleting Earth's natural resources. Almost everyone agrees this is reasonable if the consequences for the present generation are acceptable. But *why* should the planet's natural resources be preserved?

One answer is that natural resources have instrumental value. Instead of being valuable for their own sake, their value depends on what humans (and possibly other sentient beings) can do with them. While some natural resources have instrumental value for nearly all of us, others have less value. We all need stable access to safe drinking water, as no one could survive for more than a day or two without it. Note that it is not the naturalness of the water that matters. We could produce artificial freshwater from desalinated sea water, although this is an expensive and energy-intensive process.

Other natural resources are less valuable in an instrumental sense. Consider, for instance, the mineral frankamenite discovered by the Russian mineralogist V. A. Frank-Kamentsky (1915–1994). Frankamenite is an extremely rare mineral found only in the Sakha Republic in eastern Siberia, Russia. However, the fact that frankamenite is rare does not entail that it is valuable. Frankamenite is not, has never been, and will most likely never be used for anything. There are no frankamenite mines and no world market price for frankamenite. So, if we think that the value of a natural resource is determined by its instrumental value to human beings, then we have to concede that frankamenite is worthless. It would do us no harm if all frankamenite were to disappear tomorrow.

Water and frankamenite are two extremes. Many natural resources are neither highly valuable nor completely worthless in an instrumental sense. Oil might be an example of something that is valuable in an instrumental sense but that we would probably would be able to do without in the future, if we so wish (or have to).

The instrumental value of natural resources sometimes raises concerns about intergenerational fairness or justice. The popular saying that "We do not inherit the earth from our ancestors; we borrow it from our children" seems to suggest that the long-term depletion of natural resources is morally wrong because it is unfair or because future generations may need them. If we use too much of the planet's resources, there will not be enough left for future generations.

The fairness argument rests on the assumption that people in the future will value the same natural resources as we value today. It is not obvious that this assumption is true. Consider, for instance, the poisonous plant *Mandragora officinarum* (a perennial herbaceous plant). For hundreds of years, *Mandragora* was widely used for curing a variety of diseases. In those days, there were no "real" medicines available. Today you can, however, find plenty of equally good or better synthetic medicines in your local drug store. From a medical point of view, it would make no difference if *Mandragora* were to go extinct tomorrow, just as it would not matter if we were to "run out" of frankamenite. So let us imagine, contrary to the historical facts, that some new technology had been introduced twenty years ago that we knew would kill all *Mandragora* plants. Why would that have been of any concern to us if we knew no sentient being would ever need the plant again for medical (or any other) purposes?

What this example is designed to show is that the instrumental value of natural resources may vary over time. If we think that the long-term depletion of natural resources is morally wrong *because* those resources are valuable to us in an instrumental sense, then we have to concede that it would *not* be wrong to deplete quite a lot of such resources now if we knew that future generations would *not* need them.

Let us apply this line of thought to oil and other fossil fuels. For the past one hundred years or so, fossil fuels have been very important in an instrumental sense. However, it is possible that we may no longer need large quantities of fossil fuels in the future. Perhaps we will soon be able to develop cheap electric cars and airplane engines that run safely on renewable fuels. Therefore, if everyone were to need less fossil fuel in the future, it would, perhaps, not be unfair to future generations if we now consume most or all of the oil left in the ground.

The plausibility of this line of reasoning presupposes that the consumption of natural resources does not give rise to negative side effects. However, if we continue to burn large quantities of fossil fuel, we will emit huge amounts of greenhouse gases, which we know to be one of the driving forces behind climate change. That would have negative consequences for future generations. Note, however, that the argument also appeals to the *instrumental effects* of environmental degradation. The moral badness of burning fossil fuels supervenes on the fact that this will, in one way of another, lead to negative effects for people in the future. So all that matters is, it seems, how present and future people will be affected by the consumption of fossil fuels. The value of preventing any long-term depletion of natural resources is thus purely *instrumental*. It is not the natural resource itself that matters; the resource is valuable only as a means to some other end.

THE NONINSTRUMENTAL VALUE OF SUSTAINABILITY

Rather than claiming that natural resources are instrumentally valuable, some philosophers argue that nature has noninstrumental value. On this view, nature is valuable for its own sake, irrespective of its instrumental value *to us*. The moral value of a natural resource such as freshwater, oil, the mineral frankamenite, or the poisonous plant *Mandragora officinarum* does not depend on the health, welfare, or monetary values we can obtain from those resources. All these resources, and perhaps entire ecosystems, are valuable for their own sake.

The most well-known argument for this type of view is the Last Man argument proposed by Richard Routley.[5] His argument is limited to plants and other nonsentient living organisms such as bugs and ants. He does not believe that rocks and volcanoes are valuable in a noninstrumental sense.

Routley asks us to imagine a hypothetical world in which every human being except Mr. Last Man is dead, perhaps because everyone got infected by some incurable pandemic disease. Mr. Last Man knows that he will die in the near future too. He has only a few days left to live. Would Mr. Last Man, under these tragic circumstances, do anything morally wrong if he were to destroy all plants and nonsentient organisms before he dies? None of the plants and other organisms we usually value for instrumental reasons will be of any instrumental value to anyone in the future. So if we think it would be wrong to destroy nature, the best explanation seems to be that nature is not merely valuable in an instrumental sense but also in a noninstrumental sense. Here is Routley's own formulation of his argument:

> [T]he last man (or person) surviving the collapse of the world system lays about him, eliminating, as far as he can, every living thing, animal or plant (but painlessly if you like, as at the best abattoirs). What he does is quite permissible according to basic chauvinism, but on environmental grounds what he does is wrong.[6]

Is this a good argument? A possible response, is of course, that Mr. Last Man would not actually do anything wrong if he were to eliminate "as far as he can, every living thing, animal or plant." However, let us suppose, for the sake of the argument, that Routley is right about the wrongness of Mr. Last Man's actions. Would this then entail that plants and other nonsentient living organisms are valuable in a noninstrumental sense? My colleague Per Sandin and I have proposed an alternative thought experiment for testing Routley's hypothesis. This thought experiment goes as follows:

> *The Distant Nuclear Fireworks.*
> Last Man manages to escape in his spaceship just before the Earth crashes into the Sun, and he is now circling a distant planet. The on-board super computer informs him that there is some Earth-like but non-sentient life on the new planet, which will never evolve into sentient life forms since the new planet will crash into its sun within a year (which will, of course, destroy all living organisms). However, Last Man can delay this process for five years by firing a nuclear missile that will alter the planet's orbit. There are no other morally relevant aspects to consider.[7]

Does Mr. Last Man have a moral obligation to fire the nuclear missile? If Routley is right and plants and other nonsentient life really have noninstrumental value, then it would be so on all planets. Therefore, if Mr. Last Man (and no one else) could preserve some organisms for a few years, it follows that he would have a moral obligation to do so. However, this conclusion seems wrong. Intuitively, it seems that Mr. Last Man would do nothing wrong if he were to just watch the new planet slowly crash into its sun over the next couple of years.

A possible explanation of why we have different moral intuitions about Routley's original Last Man scenario and the modified spaceship version is that in the original version, Mr. Last Man is actively and aggressively destroying nature. In the second version, the effect on the ecosystem is the same, but here the planet is doomed anyway. All Mr. Last Man can do is to delay this process for a few years. So what explains our divergent intuitions about the two scenarios is Mr. Last Man's character virtues. In the original scenario, he comes about as someone who actively and aggressively kills living organisms; in the second scenario, his inaction leads to the same effect, but he is not actively and aggressively eliminating any living organisms. We can explain why Mr. Last Man's behavior is wrong (by implicitly or explicitly referring to Aristotelian virtue ethics) without ascribing any noninstrumental value to nature or ecosystems. This reveals a weakness in Routley's argument.

IMPERCEPTIBLE HARMS AND THE TRAGEDY OF THE COMMONS

Many of us find sustainability important (in an instrumental or noninstrumental sense), yet so many processes and activities in society are environmentally unsustainable. We regularly buy mineral water tapped from sources hundreds of miles away and fruit shipped for weeks across the oceans; and almost all of us fly for work or leisure throughout the year. If we value sustainability, why do we do this?

One answer could be that we do not care as much about sustainability as we say. But assume we are sincere. Most of us do care about sustainability and make some sacrifices that could benefit future generations. If so, we might explain the discrepancy between our professed values and actual behavior by appealing to a more complex mechanism than dishonesty.

Consider the following example, briefly mentioned in chapter 12. In front of the main building of your favorite university is a big, green, beautiful lawn. This lawn has significant aesthetic value appreciated by all students and faculty. Unfortunately, the grass is quite fragile. If too many students walk on it, the lawn will get muddy and lose its beauty. Every day thousands of students must decide to either (a) take a shortcut across the lawn to save a few seconds of walking time, or (b) walk around the lawn, which helps in preserving its aesthetic value. On the first day of the semester, the lawn is in perfect condition.

The first student who arrives on campus has a compelling reason to take the shortcut: by walking on the grass, the student saves time, while only imperceptibly harming the lawn. The grass will look no different after that student crosses. The second student has the same compelling reason to have crossed the lawn as the first: by taking a shortcut across the lawn, the student saves time and only imperceptibly harms the lawn. However, after five thousand students have crossed the lawn, it looks a little worn. This is compatible with the assumption that the consequences of each student's decision to cross the lawn were imperceptible. Many imperceptible consequences can together lead to some very significant and visible consequence. (Just think of climate change!)

Should student number 5,001 take the shortcut or walk around the lawn? Note that no matter what this particular student decides to do, the aesthetic value of the lawn will remain the same. Each individual crossing leads to imperceptible effects. Although the lawn is no longer in good shape, this student, too, has a compelling reason to take the shortcut, no matter how much he or she cares about preserving the lawn. After ten thousand students have crossed the lawn, it is completely destroyed.

What is interesting about this example is that even if we assume that each individual student was willing to do everything she or he could to preserve the lawn, *it was nevertheless rational for each and every student to take the shortcut.* The "harm" done to the grass by each student was imperceptible; the time saved was not imperceptible.

If we aggregate the value of the time saved by all students and compare that to the aesthetic value of the lawn, the negative aesthetic value exceeds the aggregated value of the time saved by the students. To be more precise, we stipulate that the aesthetic value of the destroyed lawn is −20,000 units, while the value of the time saved by each student is +1 unit. Clearly, the aggregated value of all the time saved, +10,000 units, does not compensate for the damage done to the lawn.

Climate change and numerous other issues related to sustainability have the same structure. It is, for instance, better for me to drive to work than to use public transportation (because public transportation is slow and uncomfortable). It also holds true that the emissions of greenhouse gases caused by a single trip make an imperceptible contribution to climate change. However, when everyone does what is rational from his or her perspective, and each act has imperceptible negative consequences, this can nevertheless lead to outcomes that are much worse for everyone.

Another mechanism that can explain why most people's actual behavior does not reflect the value they say they place on sustainability is an example from game theory known as *the tragedy of the commons* or the *n-player prisoner's dilemma* (cf. the discussion of ethical egoism in chapter 5). Imagine that the rules of your homeowner's association (HOA) stipulate that all homeowners in your neighborhood are responsible for maintaining the communal park around the corner. Every month you face a choice between either doing your fair share of the work or ducking out. The HOA cannot sanction homeowners who do not do their fair share of the work. What should you do if you are fully rational?

To simplify the structure of this problem, we assume that you have exactly two options to choose from: (a) cooperate with your neighbors to keep the park clean and tidy or (b) not cooperate. It is also helpful to think of all your neighbors as a single actor who either cooperates or does not cooperate with you. If you cooperate, and all your neighbors cooperate, then you must spend 1 hour each per month on maintaining the park, which is worth –1 unit of value to each of you. If your neighbors decide to duck out and not cooperate with you, but you clean the park yourself (you "cooperate" with yourself), then that will take you 10 hours, which is worth –10 units of value. Your neighbors get away with 0 units. However, if both you and all your neighbors decide not to cooperate and leave the park as it is, this will negatively affect the value of all your homes. The value of this loss is –5 units, measured on the same value scale as in the preceding. The numerical values we assign to the possible outcomes reflect all values and considerations you consider to be important, including the effects on the state of the park. The value assessments thus hold, *all things considered*. Consider Table 16.1. The first number of each pair denotes the value for you, and the second number denotes the value for everyone else.

What should a rational agent confronted with the matrix in Table 16.1 do? By examining the figure, we see that it is better for you not to cooperate (not do your share of the work) if the others cooperate (do their share of the work). In that case, you get 0 units of value instead of –1. It is also better for you not to cooperate if the others do not cooperate; in that case you get –5 instead of –10 units. The upshot is that it is better for you not to cooperate *no matter what the others do*. Therefore, we can expect you not to cooperate (do your part of the work) even if you care a lot about the communal park.

Let us now look at the problem from the other people's perspective. If you cooperate (do your share of the work), it is better for them not to cooperate because 0 is better for them than –1. Moreover, if you do not cooperate, it is also better for them not to cooperate, because –5 is better than –10. Therefore, it is better for them not to cooperate no matter what you do, even if they care a lot about the park.

The conclusion of all this is not that is *rational for everyone* not to do their share of the work in the park, even if *everyone* cares about it. The actual outcome will be –5, –5, which is worse for everyone than the outcome we would have reached if we had co-operated, –1, –1. The reason why this tragic situation arises is that there is no external mechanism (no police or government authority) that can *force* you do to your share of the work. It is simply too costly for you to fix the problem by yourself. If you have to do all the work to keep the park tidy, it is better for you to duck out, even if you assign a rather high value to the condition of the park.

Table 16.1 The tragedy of the commons, also known as the *n*-players prisoner's dilemma.

		Other *n* − *1* players	
		Cooperate	Do not
You	Cooperate	−1,−1	−10, 0
	Do not	0, −10	−5, −5

More generally speaking, this "tragedy of the commons" teaches us that there are situations in which the market will fail to reach acceptable solutions. Fully informed and fully rational decision makers will sometimes reach solutions to environmental problems (as well as many other types of problems) that are worse for everyone. Environmentalists appeal to this example for explaining the need for regulations that control the market forces. If we can harm the environment without paying anything for it, and as long as the homeowners in your neighborhood can refrain from doing their share of the work without being punished, we can expect rational agents to do what is best for themselves, which will typically not coincide with what is best for the environment or the common good.

REVIEW QUESTIONS

1. What were the root causes of the *Deepwater Horizon* oil spill?
2. Explain the distinction between narrow and broad sustainability, and between weak and strong sustainability.
3. What is, in your opinion, the best argument for thinking that we have a moral obligation to preserve natural resources such as oil, minerals, and living seas for future generations?
4. What claim is the Last Man argument meant to support? Is this a convincing argument?
5. What is the tragedy of the commons, and why is this concept relevant for discussions of sustainability?
6. Explain and discuss the lawn crossing example. What, if anything, does this example show?

REFERENCES AND FURTHER READINGS

Attfield, R. 1981. "The Good of Trees." *Journal of Value Inquiry* 15, no. 1: 35–54.

Benson, J. 2000. *Environmental Ethics: An Introduction with Readings.* London: Routledge.

Parfit, D. 1984. *Reasons and Persons.* Oxford, England: Oxford University Press.

Peterson, M., ed. 2015. *The Prisoner's Dilemma.* Cambridge, UK: Cambridge University Press.

Peterson, M., and P. Sandin. 2013. "The Last Man Argument Revisited." *Journal of Value Inquiry* 47, no. 1-2: 121–133.

Republic of the Marshall Islands, Office of the Maritime Administrator. August 17, 2011. *Deepwater Horizon Marine Casualty Investigation Report.* Report 2213. IMO Number 8764597.

Routley, R. 1973. "Is There a Need for a New, an Environmental, Ethic?" *Proceedings of the XV World Congress of Philosophy.* Vol. 1, 205–210. Reprinted in *Encyclopedia of Environmental Ethics and Philosophy,* edited by J. B. Callicott and R. Frodeman. Farmington Hills, MA: Macmillan Reference, 2009.

Routley, R., and V. Routley. 1980. "Human Chauvinism and Environmental Ethics?" In *Environmental Philosophy,* edited by D. S. Mannison, M. A. McRobbie, and R. Sylvan. Monograph Series No. 2. Australian National University, Department of Philosophy, Research School of Social Sciences.

Rozhdestvenskaya, I. V., and L. V. Nikishova. 1996. "The Crystal Structure of Frankamenite." *Mineralogical Magazine* 60, no. 6: 897–905.

US Environmental Protection Agency. "Sustainability and the U.S. EPA." 2015. National Academies Press website. https://www.nap.edu/read/13152/chapter/1. Accessed December 7, 2016.

Wissenburg, Marcel. *Green Liberalism: The Free and the Green Society*. London: Routledge, 2013.

World Commission on Environment and Development. 1987. *Our Common Future*. Oxford: Oxford University Press.

NOTES

1. J. Simpson, *The Oxford English Dictionary*, 3rd edition (2000–2018) Oxford: Oxford University Press, published in print and online 2000–2018).

2. Republic of the Marshall Islands, Office of the Maritime Administrator, *Deepwater Horizon Marine Casualty Investigation Report*, report 2213, issued August 17, 2011, IMO Number 8764597.

3. US Environmental Protection Agency, "Sustainability and the U.S. EPA," 2015, National Academies Press website, https://www.nap.edu/read/13152/chapter/1. Accessed December 7, 2016.

4. M. Wissenburg, *Green Liberalism: The Free and the Green Society* (London: Routledge, 2013).

5. Richard Routley, "Is There a Need for a New, an Environmental, Ethic?," in *Encyclopedia of Environmental Ethics and Philosophy*, eds. J. B. Callicott and R. Frodeman (1973; repr., Farmington Hills, MA: Macmillan Reference, 2009), 484–489.

6. Ibid. 487.

7. M. Peterson, and P. Sandin, "The Last Man Argument Revisited," *Journal of Value Inquiry* 47, no. 1–2 (2013): 130.

Professional Codes of Ethics

The National Society for Professional Engineers (NSPE)

PREAMBLE

Engineering is an important and learned profession. As members of this profession, engineers are expected to exhibit the highest standards of honesty and integrity. Engineering has a direct and vital impact on the quality of life for all people. Accordingly, the services provided by engineers require honesty, impartiality, fairness, and equity, and must be dedicated to the protection of the public health, safety, and welfare. Engineers must perform under a standard of professional behavior that requires adherence to the highest principles of ethical conduct.

I. FUNDAMENTAL CANONS

Engineers, in the fulfillment of their professional duties, shall:

1. Hold paramount the safety, health, and welfare of the public.
2. Perform services only in areas of their competence.
3. Issue public statements only in an objective and truthful manner.
4. Act for each employer or client as faithful agents or trustees.
5. Avoid deceptive acts.
6. Conduct themselves honorably, responsibly, ethically, and lawfully so as to enhance the honor, reputation, and usefulness of the profession.

II. RULES OF PRACTICE

1. Engineers shall hold paramount the safety, health, and welfare of the public.
   a. If engineers' judgment is overruled under circumstances that endanger life or property, they shall notify their employer or client and such other authority as may be appropriate.

b. Engineers shall approve only those engineering documents that are in conformity with applicable standards.

c. Engineers shall not reveal facts, data, or information without the prior consent of the client or employer except as authorized or required by law or this Code.

d. Engineers shall not permit the use of their name or associate in business ventures with any person or firm that they believe is engaged in fraudulent or dishonest enterprise.

e. Engineers shall not aid or abet the unlawful practice of engineering by a person or firm.

f. Engineers having knowledge of any alleged violation of this Code shall report thereon to appropriate professional bodies and, when relevant, also to public authorities, and cooperate with the proper authorities in furnishing such information or assistance as may be required.

2. Engineers shall perform services only in the areas of their competence.

   a. Engineers shall undertake assignments only when qualified by education or experience in the specific technical fields involved.

   b. Engineers shall not affix their signatures to any plans or documents dealing with subject matter in which they lack competence, nor to any plan or document not prepared under their direction and control.

   c. Engineers may accept assignments and assume responsibility for coordination of an entire project and sign and seal the engineering documents for the entire project, provided that each technical segment is signed and sealed only by the qualified engineers who prepared the segment.

3. Engineers shall issue public statements only in an objective and truthful manner.

   a. Engineers shall be objective and truthful in professional reports, statements, or testimony. They shall include all relevant and pertinent information in such reports, statements, or testimony, which should bear the date indicating when it was current.

   b. Engineers may express publicly technical opinions that are founded upon knowledge of the facts and competence in the subject matter.

   c. Engineers shall issue no statements, criticisms, or arguments on technical matters that are inspired or paid for by interested parties, unless they have prefaced their comments by explicitly identifying the interested parties on whose behalf they are speaking, and by revealing the existence of any interest the engineers may have in the matters.

4. Engineers shall act for each employer or client as faithful agents or trustees.

   a. Engineers shall disclose all known or potential conflicts of interest that could influence or appear to influence their judgment or the quality of their services.

   b. Engineers shall not accept compensation, financial or otherwise, from more than one party for services on the same project, or for services pertaining to the same project, unless the circumstances are fully disclosed and agreed to by all interested parties.

c. Engineers shall not solicit or accept financial or other valuable consideration, directly or indirectly, from outside agents in connection with the work for which they are responsible.

d. Engineers in public service as members, advisors, or employees of a governmental or quasi-governmental body or department shall not participate in decisions with respect to services solicited or provided by them or their organizations in private or public engineering practice.

e. Engineers shall not solicit or accept a contract from a governmental body on which a principal or officer of their organization serves as a member.

5. Engineers shall avoid deceptive acts.

a. Engineers shall not falsify their qualifications or permit misrepresentation of their or their associates' qualifications. They shall not misrepresent or exaggerate their responsibility in or for the subject matter of prior assignments. Brochures or other presentations incident to the solicitation of employment shall not misrepresent pertinent facts concerning employers, employees, associates, joint venturers, or past accomplishments.

b. Engineers shall not offer, give, solicit, or receive, either directly or indirectly, any contribution to influence the award of a contract by public authority, or which may be reasonably construed by the public as having the effect or intent of influencing the awarding of a contract. They shall not offer any gift or other valuable consideration in order to secure work. They shall not pay a commission, percentage, or brokerage fee in order to secure work, except to a bona fide employee or bona fide established commercial or marketing agencies retained by them.

III. PROFESSIONAL OBLIGATIONS

1. Engineers shall be guided in all their relations by the highest standards of honesty and integrity.

a. Engineers shall acknowledge their errors and shall not distort or alter the facts.

b. Engineers shall advise their clients or employers when they believe a project will not be successful.

c. Engineers shall not accept outside employment to the detriment of their regular work or interest. Before accepting any outside engineering employment, they will notify their employers.

d. Engineers shall not attempt to attract an engineer from another employer by false or misleading pretenses.

e. Engineers shall not promote their own interest at the expense of the dignity and integrity of the profession.

2. Engineers shall at all times strive to serve the public interest.

a. Engineers are encouraged to participate in civic affairs; career guidance for youths; and work for the advancement of the safety, health, and well-being of their community.

b. Engineers shall not complete, sign, or seal plans and/or specifications that are not in conformity with applicable engineering standards. If the client or employer insists on such unprofessional conduct, they shall notify the proper authorities and withdraw from further service on the project.

    c. Engineers are encouraged to extend public knowledge and appreciation of engineering and its achievements.

    d. Engineers are encouraged to adhere to the principles of sustainable development[1] in order to protect the environment for future generations.

3. Engineers shall avoid all conduct or practice that deceives the public.

    a. Engineers shall avoid the use of statements containing a material misrepresentation of fact or omitting a material fact.

    b. Consistent with the foregoing, engineers may advertise for recruitment of personnel.

    c. Consistent with the foregoing, engineers may prepare articles for the lay or technical press, but such articles shall not imply credit to the author for work performed by others.

4. Engineers shall not disclose, without consent, confidential information concerning the business affairs or technical processes of any present or former client or employer, or public body on which they serve.

    a. Engineers shall not, without the consent of all interested parties, promote or arrange for new employment or practice in connection with a specific project for which the engineer has gained particular and specialized knowledge.

    b. Engineers shall not, without the consent of all interested parties, participate in or represent an adversary interest in connection with a specific project or proceeding in which the engineer has gained particular specialized knowledge on behalf of a former client or employer.

5. Engineers shall not be influenced in their professional duties by conflicting interests.

    a. Engineers shall not accept financial or other considerations, including free engineering designs, from material or equipment suppliers for specifying their product.

    b. Engineers shall not accept commissions or allowances, directly or indirectly, from contractors or other parties dealing with clients or employers of the engineer in connection with work for which the engineer is responsible.

6. Engineers shall not attempt to obtain employment or advancement or professional engagements by untruthfully criticizing other engineers, or by other improper or questionable methods.

    a. Engineers shall not request, propose, or accept a commission on a contingent basis under circumstances in which their judgment may be compromised.

    b. Engineers in salaried positions shall accept part-time engineering work only to the extent consistent with policies of the employer and in accordance with ethical considerations.

    c. Engineers shall not, without consent, use equipment, supplies, laboratory, or office facilities of an employer to carry on outside private practice.

---

[1] "Sustainable development" is the challenge of meeting human needs for natural resources, industrial products, energy, food, transportation, shelter, and effective waste management while conserving and protecting environmental quality and the natural resource base essential for future development.

7. Engineers shall not attempt to injure, maliciously or falsely, directly or indirectly, the professional reputation, prospects, practice, or employment of other engineers. Engineers who believe others are guilty of unethical or illegal practice shall present such information to the proper authority for action.

   a. Engineers in private practice shall not review the work of another engineer for the same client, except with the knowledge of such engineer, or unless the connection of such engineer with the work has been terminated.

   b. Engineers in governmental, industrial, or educational employ are entitled to review and evaluate the work of other engineers when so required by their employment duties.

   c. Engineers in sales or industrial employ are entitled to make engineering comparisons of represented products with products of other suppliers.

8. Engineers shall accept personal responsibility for their professional activities, provided, however, that engineers may seek indemnification for services arising out of their practice for other than gross negligence, where the engineer's interests cannot otherwise be protected.

   a. Engineers shall conform with state registration laws in the practice of engineering.

   b. Engineers shall not use association with a nonengineer, a corporation, or partnership as a "cloak" for unethical acts.

9. Engineers shall give credit for engineering work to those to whom credit is due, and will recognize the proprietary interests of others.

   a. Engineers shall, whenever possible, name the person or persons who may be individually responsible for designs, inventions, writings, or other accomplishments.

   b. Engineers using designs supplied by a client recognize that the designs remain the property of the client and may not be duplicated by the engineer for others without express permission.

   c. Engineers, before undertaking work for others in connection with which the engineer may make improvements, plans, designs, inventions, or other records that may justify copyrights or patents, should enter into a positive agreement regarding ownership.

   d. Engineers' designs, data, records, and notes referring exclusively to an employer's work are the employer's property. The employer should indemnify the engineer for use of the information for any purpose other than the original purpose.

   e. Engineers shall continue their professional development throughout their careers and should keep current in their specialty fields by engaging in professional practice, participating in continuing education courses, reading in the technical literature, and attending professional meetings and seminars.

AS REVISED JULY 2007

By order of the United States District Court for the District of Columbia, former Section 11(c) of the NSPE Code of Ethics prohibiting competitive bidding, and all policy statements, opinions, rulings or other guidelines interpreting its scope, have been rescinded as unlawfully interfering with the legal right of engineers, protected under the antitrust laws, to provide price information to prospective clients; accordingly, nothing contained in the NSPE Code of Ethics, policy statements, opinions, rulings or other guidelines prohibits the submission of price quotations or competitive bids for engineering services at any time or in any amount.

STATEMENT BY NSPE EXECUTIVE COMMITTEE

In order to correct misunderstandings which have been indicated in some instances since the issuance of the Supreme Court decision and the entry of the Final Judgment, it is noted that in its decision of April 25, 1978, the Supreme Court of the United States declared: "The Sherman Act does not require competitive bidding." It is further noted that as made clear in the Supreme Court decision:

1. Engineers and firms may individually refuse to bid for engineering services.
2. Clients are not required to seek bids for engineering services.
3. Federal, state, and local laws governing procedures to procure engineering services are not affected, and remain in full force and effect.
4. State societies and local chapters are free to actively and aggressively seek legislation for professional selection and negotiation procedures by public agencies.
5. State registration board rules of professional conduct, including rules prohibiting competitive bidding for engineering services, are not affected and remain in full force and effect. State registration boards with authority to adopt rules of professional conduct may adopt rules governing procedures to obtain engineering services.
6. As noted by the Supreme Court, "nothing in the judgment prevents NSPE and its members from attempting to influence governmental action. . . ."

Reprinted by Permission of the National Society of Engineers (NSPE) www.nspe.org.

The Institute of Electrical and Electronics Engineers (IEEE) Code of Ethics

We, the members of the IEEE, in recognition of the importance of our technologies in affecting the quality of life throughout the world, and in accepting a personal obligation to our profession, its members and the communities we serve, do hereby commit ourselves to the highest ethical and professional conduct and agree:

1. to accept responsibility in making decisions consistent with the safety, health, and welfare of the public, and to disclose promptly factors that might endanger the public or the environment;
2. to avoid real or perceived conflicts of interest whenever possible, and to disclose them to affected parties when they do exist;
3. to be honest and realistic in stating claims or estimates based on available data;
4. to reject bribery in all its forms;
5. to improve the understanding of technology; its appropriate application, and potential consequences;

6. to maintain and improve our technical competence and to undertake techno-logical tasks for others only if qualified by training or experience, or after full disclosure of pertinent limitations;

7. to seek, accept, and offer honest criticism of technical work, to acknowledge and correct errors, and to credit properly the contributions of others;

8. to treat fairly all persons and to not engage in acts of discrimination based on race, religion, gender, disability, age, national origin, sexual orientation, gender identity, or gender expression;

9. to avoid injuring others, their property, reputation, or employment by false or malicious action;

10. to assist colleagues and co-workers in their professional development and to support them in following this code of ethics.

The Association for Computing Machinery (ACM) Code of Ethics

PREAMBLE

Commitment to ethical professional conduct is expected of every member (voting members, associate members, and student members) of the Association for Computing Machinery (ACM).

This Code, consisting of 24 imperatives formulated as statements of personal responsibility, identifies the elements of such a commitment. It contains many, but not all, issues professionals are likely to face. Section 1 outlines fundamental ethical considerations, while Section 2 addresses additional, more specific considerations of professional conduct. Statements in Section 3 pertain more specifically to individuals who have a leadership role, whether in the workplace or in a volunteer capacity such as with organizations like ACM. Principles involving compliance with this Code are given in Section 4.

The Code shall be supplemented by a set of Guidelines, which provide explanation to assist members in dealing with the various issues contained in the Code. It is expected that the Guidelines will be changed more frequently than the Code.

The Code and its supplemented Guidelines are intended to serve as a basis for ethical decision making in the conduct of professional work. Secondarily, they may serve as a basis for judging the merit of a formal complaint pertaining to violation of professional ethical standards.

It should be noted that although computing is not mentioned in the imperatives of Section 1, the Code is concerned with how these fundamental imperatives apply to one's conduct as a computing professional. These imperatives are expressed in a general form to emphasize that ethical principles which apply to computer ethics are derived from more general ethical principles.

It is understood that some words and phrases in a code of ethics are subject to varying interpretations, and that any ethical principle may conflict with other ethical principles in specific situations. Questions related to ethical conflicts can best be answered by thoughtful consideration of fundamental principles, rather than reliance on detailed regulations.

1. GENERAL MORAL IMPERATIVES
As an ACM member I will . . .

1.1 Contribute to society and human well-being.

This principle concerning the quality of life of all people affirms an obligation to protect fundamental human rights and to respect the diversity of all cultures. An essential aim of computing professionals is to minimize negative consequences of computing systems, including threats to health and safety. When designing or implementing systems, computing professionals must attempt to ensure that the products of their efforts will be used in socially responsible ways, will meet social needs, and will avoid harmful effects to health and welfare.

In addition to a safe social environment, human well-being includes a safe natural environment. Therefore, computing professionals who design and develop systems must be alert to, and make others aware of, any potential damage to the local or global environment.

1.2 Avoid harm to others.

"Harm" means injury or negative consequences, such as undesirable loss of information, loss of property, property damage, or unwanted environmental impacts. This principle prohibits use of computing technology in ways that result in harm to any of the following: users, the general public, employees, employers. Harmful actions include intentional destruction or modification of files and programs leading to serious loss of resources or unnecessary expenditure of human resources such as the time and effort required to purge systems of "computer viruses."

Well-intended actions, including those that accomplish assigned duties, may lead to harm unexpectedly. In such an event the responsible person or persons are obligated to undo or mitigate the negative consequences as much as possible. One way to avoid unintentional harm is to carefully consider potential impacts on all those affected by decisions made during design and implementation.

To minimize the possibility of indirectly harming others, computing professionals must minimize malfunctions by following generally accepted standards for system design and testing. Furthermore, it is often necessary to assess the social consequences of systems to project the likelihood of any serious harm to others. If system features are misrepresented to users, coworkers, or supervisors, the individual computing professional is responsible for any resulting injury.

In the work environment the computing professional has the additional obligation to report any signs of system dangers that might result in serious personal or social damage. If one's superiors do not act to curtail or mitigate such dangers, it may be necessary to "blow the whistle" to help correct the problem or reduce the risk. However, capricious or misguided reporting of violations can, itself, be harmful. Before reporting violations, all relevant aspects of the incident must be thoroughly assessed. In particular, the assessment of risk and responsibility must be credible. It is suggested that advice be sought from other computing professionals. See principle 2.5 regarding thorough evaluations.

1.3 Be honest and trustworthy.

Honesty is an essential component of trust. Without trust an organization cannot function effectively. The honest computing professional will not make deliberately false or deceptive claims about a system or system design, but will instead provide full disclosure of all pertinent system limitations and problems.

A computer professional has a duty to be honest about his or her own qualifications, and about any circumstances that might lead to conflicts of interest.

Membership in volunteer organizations such as ACM may at times place individuals in situations where their statements or actions could be interpreted as carrying the "weight" of a larger group of professionals. An ACM member will exercise care to not misrepresent ACM or positions and policies of ACM or any ACM units.

1.4 Be fair and take action not to discriminate.

The values of equality, tolerance, respect for others, and the principles of equal justice govern this imperative. Discrimination on the basis of race, sex, religion, age, disability, national origin, or other such factors is an explicit violation of ACM policy and will not be tolerated.

Inequities between different groups of people may result from the use or misuse of information and technology. In a fair society, all individuals would have equal opportunity to participate in, or benefit from, the use of computer resources regardless of race, sex, religion, age, disability, national origin or other such similar factors. However, these ideals do not justify unauthorized use of computer resources nor do they provide an adequate basis for violation of any other ethical imperatives of this code.

1.5 Honor property rights including copyrights and patent.

Violation of copyrights, patents, trade secrets and the terms of license agreements is prohibited by law in most circumstances. Even when software is not so protected, such violations are contrary to professional behavior. Copies of software should be made only with proper authorization. Unauthorized duplication of materials must not be condoned.

1.6 Give proper credit for intellectual property.

Computing professionals are obligated to protect the integrity of intellectual property. Specifically, one must not take credit for other's ideas or work, even in cases where the work has not been explicitly protected by copyright, patent, etc.

1.7 Respect the privacy of others.

Computing and communication technology enables the collection and exchange of personal information on a scale unprecedented in the history of civilization. Thus there is increased potential for violating the privacy of individuals and groups. It is the responsibility of professionals to maintain the privacy and integrity of data describing individuals. This includes taking precautions to ensure the accuracy of data, as well as protecting it from unauthorized access or accidental disclosure to inappropriate individuals. Furthermore, procedures must be established to allow individuals to review their records and correct inaccuracies.

This imperative implies that only the necessary amount of personal information be collected in a system, that retention and disposal periods for that information be clearly defined and enforced, and that personal information gathered for a specific purpose not be used for other purposes without consent of the individual(s). These principles apply to electronic communications, including electronic mail, and prohibit procedures that capture or monitor electronic user data, including messages, without the permission of users or bona fide authorization related to system operation and maintenance. User data observed during the normal duties of system operation and maintenance must be treated with strictest confidentiality, except in cases where it is evidence for the violation of law, organizational regulations, or this Code. In these cases, the nature or contents of that information must be disclosed only to proper authorities.

1.8 Honor confidentiality.

The principle of honesty extends to issues of confidentiality of information whenever one has made an explicit promise to honor confidentiality or, implicitly, when private information not directly related to the performance of one's duties becomes available. The ethical concern is to respect all obligations of confidentiality to employers, clients, and users unless discharged from such obligations by requirements of the law or other principles of this Code.

2. MORE SPECIFIC PROFESSIONAL RESPONSIBILITIES
As an ACM computing professional I will . . .

2.1 Strive to achieve the highest quality, effectiveness and dignity in both the process and products of professional work.

Excellence is perhaps the most important obligation of a professional. The computing professional must strive to achieve quality and to be cognizant of the serious negative consequences that may result from poor quality in a system.

2.2 Acquire and maintain professional competence.

Excellence depends on individuals who take responsibility for acquiring and maintaining professional competence. A professional must participate in setting standards for appropriate levels of competence, and strive to achieve those standards. Upgrading technical knowledge and competence can be achieved in several ways: doing independent study; attending seminars, conferences, or courses; and being involved in professional organizations.

2.3 Know and respect existing laws pertaining to professional work.

ACM members must obey existing local, state, province, national, and international laws unless there is a compelling ethical basis not to do so. Policies and procedures of the organizations in which one participates must also be obeyed. But compliance must be balanced with the recognition that sometimes existing laws and rules may be immoral or inappropriate and, therefore, must be challenged. Violation of a law or regulation may be ethical when that law or rule has inadequate moral basis or when it conflicts with another law judged to be more important. If one decides to violate a law or rule because it is viewed as unethical, or for any other reason, one must fully accept responsibility for one's actions and for the consequences.

2.4 Accept and provide appropriate professional review.

Quality professional work, especially in the computing profession, depends on professional reviewing and critiquing. Whenever appropriate, individual members should seek and utilize peer review as well as provide critical review of the work of others.

2.5 Give comprehensive and thorough evaluations of computer systems and their impacts, including analysis of possible risks.

Computer professionals must strive to be perceptive, thorough, and objective when evaluating, recommending, and presenting system descriptions and alternatives. Computer professionals are in a position of special trust, and therefore have a special responsibility to provide objective, credible evaluations to employers, clients, users, and the public. When providing evaluations the professional must also identify any relevant conflicts of interest, as stated in imperative 1.3.

As noted in the discussion of principle 1.2 on avoiding harm, any signs of danger from systems must be reported to those who have opportunity and/or responsibility to resolve them. See the guidelines for imperative 1.2 for more details concerning harm, including the reporting of professional violations.

2.6 Honor contracts, agreements, and assigned responsibilities.

Honoring one's commitments is a matter of integrity and honesty. For the computer professional this includes ensuring that system elements perform as intended. Also, when one contracts for work with another party, one has an obligation to keep that party properly informed about progress toward completing that work.

A computing professional has a responsibility to request a change in any assignment that he or she feels cannot be completed as defined. Only after serious consideration and with full disclosure of risks and concerns to the employer or client, should one accept the assignment. The major underlying principle here is the obligation to accept personal accountability for professional work. On some occasions other ethical principles may take greater priority.

A judgment that a specific assignment should not be performed may not be accepted. Having clearly identified one's concerns and reasons for that judgment, but failing to procure a change in that assignment, one may yet be obligated, by contract or by law, to proceed as directed. The computing professional's ethical judgment should be the final guide in deciding whether or not to proceed. Regardless of the decision, one must accept the responsibility for the consequences.

However, performing assignments "against one's own judgment" does not relieve the professional of responsibility for any negative consequences.

2.7 Improve public understanding of computing and its consequences.

Computing professionals have a responsibility to share technical knowledge with the public by encouraging understanding of computing, including the impacts of computer systems and their limitations. This imperative implies an obligation to counter any false views related to computing.

2.8 Access computing and communication resources only when authorized to do so.

Theft or destruction of tangible and electronic property is prohibited by imperative 1.2—"Avoid harm to others." Trespassing and unauthorized use of a computer or communication system is addressed by this imperative. Trespassing includes accessing communication networks and computer systems, or accounts and/or files associated with those systems, without explicit authorization to do so. Individuals and organizations have the right to restrict access to their systems so long as they do not violate the discrimination principle (see 1.4). No one should enter or use another's computer system, software, or data files without permission. One must always have appropriate approval before using system resources, including communication ports, file space, other system peripherals, and computer time.

3. ORGANIZATIONAL LEADERSHIP IMPERATIVES
As an ACM member and an organizational leader, I will . . .

3.1 Articulate social responsibilities of members of an organizational unit and encourage full acceptance of those responsibilities.

Because organizations of all kinds have impacts on the public, they must accept responsibilities to society. Organizational procedures and attitudes oriented toward quality and the welfare of society will reduce harm to members of the public, thereby serving public interest and fulfilling social responsibility. Therefore, organizational leaders must encourage full participation in meeting social responsibilities as well as quality performance.

3.2 Manage personnel and resources to design and build information systems that enhance the quality of working life.

Organizational leaders are responsible for ensuring that computer systems enhance, not degrade, the quality of working life. When implementing a computer system, organizations must consider the personal and professional development, physical safety, and human dignity of all workers. Appropriate human-computer ergonomic standards should be considered in system design and in the workplace.

3.3 Acknowledge and support proper and authorized uses of an organization's computing and communication resources.

Because computer systems can become tools to harm as well as to benefit an organization, the leadership has the responsibility to clearly define appropriate and inappropriate uses of organizational computing resources. While the number and scope of such rules should be minimal, they should be fully enforced when established.

3.4 Ensure that users and those who will be affected by a system have their needs clearly articulated during the assessment and design of requirements; later the system must be validated to meet requirements.

Current system users, potential users and other persons whose lives may be affected by a system must have their needs assessed and incorporated in the statement of requirements. System validation should ensure compliance with those requirements.

3.5 Articulate and support policies that protect the dignity of users and others affected by a computing system.

Designing or implementing systems that deliberately or inadvertently demean individuals or groups is ethically unacceptable. Computer professionals who are in decision making positions should verify that systems are designed and implemented to protect personal privacy and enhance personal dignity.

3.6 Create opportunities for members of the organization to learn the principles and limitations of computer systems.

This complements the imperative on public understanding (2.7). Educational opportunities are essential to facilitate optimal participation of all organizational members. Opportunities must be available to all members to help them improve their knowledge and skills in computing, including courses that familiarize them with the consequences and limitations of particular types of systems. In particular, professionals must be made aware of the dangers of building systems around oversimplified models, the improbability of anticipating and designing for every possible operating condition, and other issues related to the complexity of this profession.

4. COMPLIANCE WITH THE CODE

As an ACM member I will . . .

4.1 Uphold and promote the principles of this Code.

The future of the computing profession depends on both technical and ethical excellence. Not only is it important for ACM computing professionals to adhere to the principles expressed in this Code, each member should encourage and support adherence by other members.

4.2 Treat violations of this code as inconsistent with membership in the ACM.

Adherence of professionals to a code of ethics is largely a voluntary matter. However, if a member does not follow this code by engaging in gross misconduct, membership in ACM may be terminated.

Essay Questions

This appendix contains seven essay questions. Each question focuses on a case in which the moral obligations of engineers need to be clarified. Pick a case you find interesting and write an essay of approximately three thousand words. You are expected to use the Internet and other appropriate resources for gathering the morally relevant facts of your case; the information you need is not stated in the question. Your essay should address the following issues:

1. What are the morally relevant facts of the case?
   a. What happened?
   b. Why did it happen?
   c. Who made the key decision(s)?
   d. Who was affected?
   e. What were the long-term consequences?

2. What is the most important *moral* issue raised by the case? Try to formulate the moral question in a single sentence.
3. Define the key terms and clarify any unresolved conceptual issues. (Example: Did anyone receive a "bribe" or was it merely a "complimentary gift"?)
4. Apply the relevant code(s) of ethics. Do the code(s) issue clear and coherent verdicts? If not, why not?
5. How robust is the moral verdict issued by the code(s)? Check this by applying the ethical theories discussed in chapters 5 and 6. Do all ethical theories yield the same verdict? If not, develop a strategy for managing this moral uncertainty.
6. Formulate a moral conclusion about the case and discuss potential objections to your conclusion.

Case I: What was the role of engineer James Liang in the Volkswagen emission scandal that surfaced in 2015? Can his actions be justified or excused? Search for reliable information about Mr. Liang's actions on the Internet or by using other appropriate resources.

Case II: What was the most important moral issue raised by the Flint, Michigan, drinking water crises in 2014, and how should it have been managed? Search for reliable information about the crisis on the Internet or by using other appropriate resources.

Case III: On November 7, 1940, the Tacoma Narrows Suspension Bridge began to oscillate heavily. After about one hour, it collapsed. Othmar Ammann, a member of the team that investigated the accident, famously concluded that "The Tacoma Narrows bridge failure has given us invaluable information [about bridge design] . . . we must rely largely on judgement and if, as a result, errors, or failures occur, we must accept them as a price for human progress." Do you agree? Search for reliable information about the Tacoma bridge failure on the Internet or by using other appropriate resources.

Case IV: On March 17, 2018, several newspapers reported that at least fifty million Facebook profiles had been secretly harvested for data by Cambridge Analytica. Was Mark Zuckerberg, CEO of Facebook, personally responsible for this breach of trust? If not, can we hold corporate entities morally responsible for their actions? Search for reliable information about the Cambridge Analytica scandal on the Internet or by using other appropriate resources.

Case V: The Banqiao Dam in Henan, China, was completed by Soviet engineers in 1952. On August 9, 1975, unusually high water levels in combination with poorly repaired cracks led to the collapse of the dam. This is the deadliest structural disaster in world history. Approximately two hundred thousand people died and about eleven million people were forced to abandon their homes. Analyze the role of hydrologist Chen Xing in the Banqiao Dam disaster and discuss what, if anything, we can learn about environmental justice from this tragic event. Search for reliable information on the Internet or by using other appropriate resources.

Case VI: What happened in the Union Carbide pesticide plant in Bhopal, India, on December 2, 1984, and what can we learn from it today? Search for reliable information about the accident on the Internet or by using other appropriate resources.

Case VII: What, if anything, can we learn about the moral responsibilities of engineers from the accident in reactor number 2 in the Three Mile Island Nuclear Engineering Station on March 28, 1979? Search for reliable information about the accident on the Internet or by using other appropriate resources.

Act utilitarianism—The ethical theory holding that a particular act is right just in case the consequences of that act are optimal.

Actant—According to French sociologist Bruno Latour, an actant is a technological artifact that performs actions. Example: a speed bump that slows down cars.

Actor-Network Theory—A theory proposed by French sociologist Bruno Latour, according to which no sharp distinction can be made between objects (the external world) and subjects (human beings).

Anthropocentric ethics—The view that the primary concern of ethics is the interests, needs, or well-being of human beings, not the interests, needs, or well-being of animals, plants, or ecosystems.

Artifact—A man-made object or thing. Example: a speed bump.

Aspirational ethics—Ethical principles that go beyond the bare minimum required for avoiding wrongdoing. Example: An engineer promotes the welfare of society by working for Engineers Without Borders in his or her free time.

Association for Computer Machinery (ACM)—The world's largest society for computing professionals, with thousands of members around the world.

Casuistry—A case-based method of ethical reasoning, holding that no general moral principles are required for reaching warranted conclusions about real-world cases.

Categorical imperative—A moral command in Kant's ethical theory that is valid under all circumstances and in all situations, regardless of the agent's wishes or desires. Example: "Act only according to that maxim whereby you can, at the same time, will that it should become a universal law."

Co-constructivism—An intermediate position between technological determinism and social constructivism that attributes the power to change history to technology itself as well as to a multitude of social forces: socioeconomic, cultural, ideological, and political factors.

Code of ethics—A set of moral rules for managing ethical problems within a specific (professional) domain.

Coherence (in ethics)—The property a set of moral judgments has when they support and explain each other well.

Conceptual claim—A claim that is true or false in virtue of its meaning. Example: No triangle has more than four sides.

Conflict of interest—A set of circumstances that creates a risk that someone's (professional) judgment will be unduly influenced by an inappropriate interest or consideration. A conflict of interest can be actual, potential, or merely apparent.

Courage—The virtue of acting in accordance with one's (moral) convictions or beliefs.

Criterion of rightness—Necessary and sufficient conditions that separate morally right acts from wrong ones.

Critical attitude to technology—The view that technology as such is neither good nor bad, although some particular technologies may sometimes warrant criticism; cf. technological optimism and pessimism.

Dichlorodiphenyltrichloroethane (DDT)—A highly toxic pesticide, which many experts falsely deemed to be safe for humans when it was first discovered. It was banned by the EPA in 1972.

Doctrine of the mean—The claim that a virtuous agent should strive for the desirable middle between deficiency and excess.

Domain-specific moral principle—A moral principle that is applicable only to moral problems within a specific domain of professional ethics.

Duty ethics—An ethical theory according to which an act's moral rightness or wrongness depends on the intention with which it is performed rather than its consequences or the agent's act dispositions.

Egalitarianism—The view that well-being, primary social goods, or other bearers of value ought to be distributed equally in society.

Ethical egoism—The view that it is morally right for an agent to do whatever produces the best consequences for that agent.

Ethical theory—A general claim about what makes morally right acts right and wrong ones wrong. Examples: Utilitarianism, duty ethics, virtue ethics, and rights-based theories.

Ethics—Moral principles, values, virtues, or other considerations that govern our behavior toward other persons or morally relevant entities.

Eudaimonia—A technical term in virtue ethics that is often translated as human flourishing or happiness.

Existentialism—A philosophical theory developed by French philosopher Jean-Paul Sartre and others that emphasizes the individual's freedom and responsibility to make decisions by exercising free will.

Expected value—The probability that some outcome will occur multiplied by the value of that outcome.

Factual claim—A claim whose truth or falsity depends on what facts of the world obtain or do not obtain. Example: Many engineering students take courses in engineering ethics.

Foam shedding—The process in which insulating foam falls off from a space shuttle during the launch sequence; the cause of the Columbia disaster in 2003.

Ford Pinto—A small and cheap automobile manufactured by Ford between 1971 and 1980 that was infamous for its poor safety record caused by a poorly designed fuel tank that was prone to explode in rear-end collisions.

Foreign Corrupt Practices Act (FCPA)—This law prohibits US citizens and corporations to make payments to foreign officials to assist in obtaining or retaining business, even if such payments are legal in the country in which they are made.

Global Manhattan—A thought experiment in which the entire planet is covered with highways, skyscrapers, airports, factories, and a few artificial parks for recreation.

Golden rule, the—The moral principle according to which you should "do unto others as you would have them do unto you."

Hedonic pricing—A method for assigning monetary values to non-market goods, in particular the environment.

Hume's Law—David Hume's (1711–1776) claim that we cannot derive an "ought" from an "is": every valid inference to a moral conclusion requires at least one (nonvacuous) moral premise.

Hybrid agent—A moral agent comprising human as well as technological parts.

Hypothetical imperative—A moral command in Kant's ethical theory that is valid only if the agent has certain wishes or desires. Example: "If you wish to eat good pasta, then you should dine in Little Italy in New York."

Imperfect duty—A duty that, according to Kant, does not have to be fulfilled under all circumstances.

Informed consent—The moral principle holding that it is morally permissible to do something to another person, or impose a risk on another person, only if the person affected by the act has been properly informed about the possible consequences and consented to them.

Institute of Electrical and Electronics Engineers (IEEE)—The world's largest professional organization for engineers with hundreds thousands of members around the world.

Instrumental value—Something is valuable in an instrumental sense just in case it is valuable as a means to an end, rather than valuable for its own sake. Example: money.

Intrinsic value—Something is valuable in an intrinsic sense just in case it is valuable as a means to an end, rather than as a means to an end. Example: happiness.

Legal positivism—The view that law and morality are entirely distinct entities. Laws are social constructions; and we cannot infer anything about what is, or should be, legally permitted from claims about what is morally right or wrong.

Macroethics—The investigation of moral issues related to large-scale societal problems, such as global warming.

Maxim—A *maxim* is the rule that governs the intention with which one is acting. It can often be formulated by stating the agent's reason for doing something.

Microethics—The investigation of moral issues that concern the behavior of individuals or small groups of people, for example, moral issues related to conflicts of interest in one's workplace.

Mixing theory of labor—The view that one becomes the owner of something if one mixes one's labor with something that is not owned by anyone, while leaving enough left for others.

Moral claim—A claim that expresses a moral judgment. Example: Engineers shall hold paramount the safety, health, and welfare of the public.

Moral dilemma—In a narrow, academic sense, a moral dilemma is a situation in which all alternatives open to the agent are morally wrong. Such moral dilemmas are by definition irresolvable. In ordinary, nonacademic contexts, a moral dilemma is a difficult moral choice situation, which need not always be irresolvable.

Moral principle—A claim about how one ought to behave.

Moral realism—The metaethical view that every moral statement is either true or false, in a sense that is independent of our feelings, attitudes, social conventions, and other similar social constructions.

Moral relativism—The metaethical view that moral statements are true or false relative to some cultural tradition, religious conviction, or subjective opinion.

Morals—Ethical principles, values, virtues, or other considerations that govern our behavior toward other persons or ethically relevant entities.

Narrow sustainability—The view that sustainability can only be achieved by eliminating all forms of significant, long-term depletion of natural resources.

National Society of Professional Engineers (NSPE)—A learned society for engineers that addresses the professional concerns of licensed engineers (called Professional Engineers, or PEs for short) across all engineering disciplines.

Natural law theory—The view that moral concerns determine what is, or should be, legally permissible or impermissible.

Normalization of deviance—The process in which a technical error is accepted as normal, even though the technological system is not working as it should.

Nuremberg code—An influential set of ethical principles governing research on human subjects adopted in the wake of World War II.

Particularism—The view that ethical problems can and must be resolved without invoking any moral principle, only particular judgments about individual cases.

Perfect duty—A duty that, according to Kant, has to be fulfilled under all circumstances.

Precautionary principle—The principle holding that reasonable precautionary measures ought to be taken to safeguard against uncertain but nonnegligible threats.

Preventive ethics—Moral principles that seek to prevent accidents and other types of problems from arising.

Principlism—A method of ethical reasoning in which a set of domain-specific moral principles are balanced against each other for reaching a warranted conclusion about some real-world case.

Prioritarianism—Prioritarians believe that benefits to those who are worse off count for more than benefits to those who are better off. Well-being has a decreasing marginal moral value, just like most people have a decreasing marginal utility for money.

Prisoner's dilemma—A type of situation in which rational and fully informed agents act in accordance with their self-interest that is suboptimal from each agent's own point of view.

Problem of many hands—A situation in which there is a "gap" in the distribution of responsibility within a group of agents. The total responsibility assigned to the individual members of the group may, for instance, not accurately reflect the magnitude of a disaster.

Professional Engineer (PE)—In the United States, a Professional Engineer is someone who has obtained a license to practice engineering by taking written tests and gaining some work experience. Only Professional Engineers can become members of the National Society of Professional Engineers (NSPE).

Prohibitive ethics—Moral principles that seek to prohibit certain types of actions, for example, cheating and bribery.

Proper engineering decision (PED)—A PED is a decision that requires technical expertise and may significantly affect the health, safety, and welfare of others, or has the potential to violate the standards of an engineering code of ethics in other ways.

Proper management decision (PMD)—A PMD is a decision that affects the performance of the organization but does not require any technical expertise and does not significantly affect the health, safety, and welfare of others, nor has any potential to violate the standards of any engineering code of ethics.

Prudence—The virtue of being cautious and exercising good judgment.

Quality Adjusted Life Years—A method for assigning monetary values to non-market goods in the healthcare sector.

Right to be forgotten—In the European Union, individuals have the right to ask search engines to remove links with personal information about them; and under certain conditions, the search engines are obliged to comply.

Risk—The term risk has different meanings in different contexts. According to the *engineering definition of risk*, the risk of some unwanted event is the product of the probability that the event will occur and the value of the harm caused by the event, measured in whatever unit deemed appropriate.

Rule utilitarianism—An ethical theory holding that we ought to act per a set of rules of that would lead to optimal consequences if they were to be accepted by an overwhelming majority of people in society.

Social experiment—An experiment in which a large group of individuals in society participate as test subjects.

Socially constructed right—The view that rights are created by society. On this view, we do not have any rights merely in virtue of being moral agents (or patients).

Technê—An ancient Greek term often translated as the craft or art to build or make things.

Techno-fix—A technological solution to a social problem.

Technological determinism—According to Heilbroner, this is the view that "there is a fixed sequence to technological development and therefore a necessitous path over which technologically developing societies must travel."

Technological mediation—The claim that technological artifacts sometimes change or enhance our perception of the world or our actions.

Technological optimism—The view that technological progress has mostly improved our living conditions and that further technological advancements would make us even better off.

Technological pessimism—The view that technological progress has no or little value and that we are therefore no better off than we would have been without modern technology.

Technology assessment—A method for predicting and assessing the consequences of a new technology.

Temperance—The virtue of moderation or self-restraint.

Tragedy of the commons—This is a version of the prisoner's dilemma with more than two agents.

Travel cost method—A method for assigning monetary values to non-market goods, in particular, the environment.

Trichlorethylene—A clear, nonflammable liquid commonly used as a solvent. For many years, it remained uncertain whether trichlorethylene is a human carcinogen. In 2011, the US National Toxicology Program concluded that trichloroethylene can be "reasonably anticipated to be a human carcinogen."

Trolley Problem—A moral choice situation in which you have to choose between killing one person, which will prevent the death of five others, or let the five die.

Utilitarianism—The ethical theory holding that an act is right just in case it brings about the greatest sum total of pleasure or well-being for everyone affected by the act.

Virtue—An act disposition or character trait that is stable over time, in particular, ones that characterize morally excellent individuals.

Whistle-blowing—The act of breaking with protocol to bypass the ordinary chain of command by, for example, contacting the press (external whistle-blowing) or the supervisor's supervisor (internal whistle-blowing) to reveal serious moral or legal wrongdoing.

Figures, notes, and tables are denoted by f, n, and t, respectively.